Making Less Garbage

A Planning Guide for Communities

Bette K. Fishbein and Caroline Gelb

INFORM, Inc.
381 Park Avenue South
New York, NY 10016-8806
Tel 212 689-4040
Fax 212 447-0689

Library of Congress Cataloging-in-Publication Data

Fishbein, Bette K.
    Making less garbage : a planning guide for communities / Bette K. Fishbein,
and Caroline Gelb.
       p.   cm.
    Includes bibliographical references and index.
    ISBN 0-918780-58-6 : $30.00
    1. Source reduction (Waste management) 2. Source reduction (Waste manage-
ment)–United States. I. Gelb, Caroline. II. Title.
TD793.95.F57   1992
363.72'8--dc20   92-27072
                                CIP

INFORM, Inc., founded in 1974, is a nonprofit research organization that identifies and reports on practical actions for the protection and conservation of natural resources and public health. INFORM's research is published in books, abstracts, newsletters, and articles. Its work is supported by contributions from individuals and corporations and by grants from over 40 foundations.

Third printing

Printed on recycled paper

Table of Contents

Tables and Figures

Acknowledgments

We wish to thank many people for their contributions to this report. First, we are grateful to the dozens of people running source reduction programs throughout the country who provided us with examples of what can be done that we have included in this book. We extend special thanks to the following people who provided support and insightful review comments: Ken Brown, Waste Reduction Manager, Minnesota Office of Waste Management; Lisa Fernandez, Waste Prevention Specialist, New York City Department of Sanitation; Ellen Harrison, Associate Director, Cornell Waste Management Institute; John Schall, Visiting Fellow, Yale University Program on Solid Waste Policy and Project Director, New York City Solid Waste Management Plan, Tellus Institute; and Carl Woestwin, Waste Reduction Planner, Seattle Solid Waste Utility.

A particular note of appreciation is due to Nancy Lilienthal, formerly Director of INFORM's Chemical Hazards Prevention Program, for writing the chapter on reducing toxics in the waste stream. We also thank the other staff members who helped develop and complete this book. Joanna D. Underwood, President of INFORM, provided support and valuable suggestions. Sibyl R. Golden, Director of Research and Publications, supplied endless organizational and editorial direction that helped create a useable document. David Saphire, Research Associate, made significant contributions as an informational resource and reviewer. Thanks also go to Diana Weyne for copy editing; to Elisa Last for design and layout; and to Sharene Azimi for proofreading.

Finally, we thank the organizations whose support made this guide possible: Region II of the US Environmental Protection Agency, and, in addition, the Mary Reynolds Babcock Foundation, the Robert Sterling Clark Foundation, Inc., the Geraldine R. Dodge Foundation, and the Victoria Foundation, Inc. The support provided by the EPA does not signify that the contents necessarily reflect the views and policies of the US EPA.

While we could not have gathered and analyzed the information in this book without the assistance of all these people, the contents, findings, and conclusions are the sole responsibility of INFORM.

Preface

Every day, every one of us generates garbage — in our homes, in our workplaces, in our schools, and in our institutions. From that simple fact springs an equally simple-sounding solution to our nation's growing solid waste problem: we could each generate less garbage.

Source reduction, as this strategy is known, tops the list of solid waste management options adopted by the US Environmental Protection Agency, followed by recycling (including composting), incineration, and landfilling. It decreases the amount (and toxicity) of waste that must be managed; in so doing, it also reduces the growing costs of collection, recycling, and disposal systems and helps alleviate the political conflicts that proposals for creating these systems often engender.

Most states and localities have also endorsed source reduction and, increasingly, communities are being required to develop comprehensive solid waste management plans that include a source reduction component. Yet, until recently, source reduction, hailed in principle, has been neglected in practice. Why? Because solid waste planners, while recognizing the benefits of source reduction, have not had a clear picture of how to make it happen.

For the hundreds of source reduction planners across the country, as well as the government and community officials wanting to see less garbage, this manual offers clear guidance for designing municipal source reduction initiatives. *Making Less Garbage: A Planning Guide for Communities* provides the information they need to develop and implement successful programs to reduce both the amount and toxicity of waste. It discusses the steps for creating an effective source reduction program, from defining policies, goals, and measurement systems to establishing an administrative structure and adequate budget. It describes specific source reduction actions that are already being carried out around the country. And it provides a planning checklist that solid waste officials can use as an outline for developing their own source reduction programs.

The potential for source reduction is enormous. Through it, we help preserve our country's natural resources and, equally importantly, we help make the United States — to date, one of the highest waste-generating nations in the world — an example of change. We demonstrate our recognition that this planet has limits. We show our willingness to alter our practices and take real steps towards sustainability. It is INFORM's hope that *Making Less Garbage* will be the tool that will help government planners achieve this potential.

The dozens of successful source reduction initiatives described in this guide come from every sector of US society: state and local governments; businesses of all sizes; nonprofit organizations; public institutions such as schools, hospitals, and parks; citizens' groups; individual consumers; and nonprofit organizations. Virtually every person and every organization can play a role in reducing waste — and become part of the solution to our country's (and the world's) garbage crisis.

Joanna D. Underwood
President
INFORM

PART I   THE ESSENTIALS OF SOURCE REDUCTION
          PLANNING

▓ Chapter 1 Introduction

A solid waste crisis is confronting the United States: even as we are producing ever-increasing amounts of garbage, our ability and willingness to bear the environmental and economic costs of managing and disposing of this garbage are decreasing.

From 1960 to 1988, United States garbage more than doubled, rising from 88 to 180 million tons per year. The average US resident is producing 50 percent more waste now than 30 years ago — 4.0 pounds per person per day, up from 2.7 pounds per person per day, and projected to grow to 4.9 pounds per person per day by 2010.[1] This per capita rate exceeds that of most other industrialized countries with comparable standards of living. At the same time, the US population has also dramatically increased: from 150 million people in 1950 to 250 million in 1990.

The United States not only has more garbage now, but that waste contains more toxic chemicals. Chemical production has soared, and with it a multitude of consumer products (such as paints, insecticides, batteries, inks, and solvents) that may contain toxic constituents. When these products end up in the municipal solid waste stream, environmentally acceptable disposal becomes more difficult and more costly.

The municipal solid waste stream consists of the materials thrown out by the residential, commercial, and institutional sectors. These materials include such items as paper waste, food, packaging, yard waste, clothing, and appliances. Industrial wastes such as corrugated boxes, wood pallets, lunchroom wastes, and office paper waste are included in municipal solid waste, but industrial process wastes, sludge, and ash are not.

Historically, solid waste management evolved in an ad-hoc fashion. Garbage was dumped in the ocean or on open land, with or without environmental controls, or it was burned in the open or in incinerators. When disposal facilities were filled up, there was the option of moving to virgin territories to build new ones. In the United States, with its vast open spaces, the philosophy of the expanding frontier prevailed. There was little concept of limits — of running out of places to put our trash or the need to plan ahead.

Developments over the past two decades have changed this. The United States is, for the first time, becoming aware of limits. Parts of this country are running out of places to put their garbage that are environmentally and politically acceptable. Growing sensitivity to environmental impacts is feeding public antipathy to incinerators and landfills. As heavily populated states such as New York and New Jersey export their waste to states across the country, some receiving states (including Iowa, Indiana, Michigan, Nevada, Ohio, and Washington) are moving to close their borders to out-of-state waste or to impose extra fees on such shipments. While such state approaches have been held unconstitutional (because of interstate commerce provisions), congressional action may substantially limit interstate export options in the future. As of mid-1992, 20 bills were pending in the US Congress to permit states to restrict interstate shipments of waste.

This report explores a method that can play a vital role in alleviating this country's solid waste problem, one that can not just halt the increase in per capita garbage generation but ultimately reverse it. That method is source reduction — simply put, creating less waste in the first place.

❖ Source Reduction: Hailed in Principle, Limited in Practice

Source reduction — reducing the amount and/or toxicity of waste actually generated — offers promising environmental and economic opportunities. It means that communities need to collect, process, and dispose of less waste, thereby reducing both their waste management costs and the environmental impacts of these activities. It requires no waste management facilities. The case for source reduction is the case for prevention rather than remediation: the old saying that an ounce of prevention is worth a pound of cure is as true for garbage as it is for health.

In fact, source reduction tops the list of solid waste management options adopted by the US Environmental Protection Agency and endorsed by most states and localities: [2]

1. Source reduction and reuse
2. Recycling of materials, including composting of yard and some food waste
3. Waste combustion
4. Landfilling

The EPA's list is a hierarchy; that is, it does not just present options but establishes an order of priorities, placing source reduction before recycling, and recycling before disposal. Despite this statement of principle, source reduction remains neglected in practice, as dramatized by the ever-growing per capita waste generation rates.

Planning for Source Reduction

In attempting to deal with the US solid waste problem, the 1976 Resource Conservation and Recovery Act (RCRA) required states to draw up solid waste plans. More than three-quarters of the 50 states (38 in all), in response, either have mandated that localities within their borders develop comprehensive long-term solid waste plans or have put incentives in place to encourage such planning (see Table 3-1). These local plans will shape solid waste policy for the future, since local governments have the primary responsibility for managing solid waste. In addition, two smaller states (Delaware and Rhode Island) are developing state-level plans, and the District of Columbia is developing its own plan.

The state requirements are causing many communities to confront, for the first time, the issue of solid waste planning in general, and source reduction in particular. While generally endorsing the EPA priorities, most of the plans that have been developed so far fail to include specific initiatives to accomplish source reduction or to allocate the resources needed to carry them out.

The Source Reduction Challenge

Why has the preventive strategy agreed by all to be the top priority been treated, in fact, like a poor relative in this planning process? The answer lies in the experience of most solid waste managers and the nature of source reduction itself.

Local solid waste planners, engineers, and managers know the steps involved in designing and operating landfills, incinerators, and collection systems. They know that recycling involves collection, separation, and marketing of materials. However, while acknowledging source reduction as the most important priority, most have only a vague idea of how to get businesses and citizens to produce less waste. Most are uncertain about what steps municipalities can take on their own and what steps can most effectively be taken at the state or federal level.

Source reduction can also seem in many ways like a "non-event"; it is less tangible than disposal or recycling. While building incinerators, landfills, and recycling centers involves definable activities, preventing the generation of garbage implies facilities that are not built and materials that are not collected, not marketed, and not sold. Focusing on these issues has been more difficult for planners accustomed to dealing with more tangible materials and activities.

Furthermore, there is an underlying concern that source reduction may be anti-prosperity or anti-progress. There is some basis to this concern, since we would clearly generate less waste if we consumed and produced less: waste generation declines in recessions. The challenge, however, is to sustain economic growth while being less wasteful — by increasing efficiency and creating new repair and other jobs.

❖ INFORM's Guide: How to Make Source Reduction Happen

The good news that has emerged from the INFORM research that has resulted in this planning guide is that source reduction can be a positive energetic program initiative that municipalities can make happen. It has the potential not only to reduce the growing costs of collection and disposal systems, making funds available for other purposes, but also to eliminate the political conflicts that proposals for creating these systems often engender.

INFORM created this guide to identify key components of local source reduction plans — organizational features and individual projects needed to make source reduction happen. The project originated in response to a specific situation in New York State (a 1988 requirement that each of the 63 state solid waste planning districts develop and obtain state approval for a solid waste management plan that included strategies for meeting state goals of 8 to 10 percent source reduction and 40 to 42 percent recycling by 1997). However, the findings are relevant to all solid waste planners around the country, whether they are developing source reduction plans or critiquing those prepared by others. The information about specific initiatives can also be helpful to businesses and community groups that want to develop their own source reduction programs and/or work with local governments.

Methodology of the Study

To develop this guide, INFORM looked around the country for successful source reduction efforts, whether they were undertaken by local governments, state governments, private industry, institutions, or nonprofit organizations. INFORM contacted solid waste officials in all 50 states and the District of Columbia to determine whether they had established source reduction goals and plans, and conducted in-depth telephone interviews with representatives from 27 states[3] to identify specific strategies that had actually been implemented. These state-level representatives provided leads to local solid waste planners and to businesses, institutions, and citizens' groups that had undertaken source reduction activities.

Source reduction is still in its infancy, so little has been published on the topic. However, INFORM did obtain some information about source reduction initiatives from journals, books, and conference presentations.

Organization of this Guide

This source reduction planning guide is divided into three parts. Part I, "The Essentials of Source Reduction Planning," discusses the US solid waste problem and the benefits of source reduction; describes how communities can establish source reduction policies, goals, and measurement systems; and outlines the administrative structure and budget requirements for an effective source reduction program.

Part II, "Source Reduction Initiatives," describes dozens of specific activities to reduce the amount and/or toxicity of solid waste that are being carried out around the country. These activities include:

- Government source reduction programs (procurement and operations)
- Institutional source reduction programs (in prisons, hospitals, and schools)
- Government programs to stimulate source reduction action (technical assistance, backyard composting and leave-on-lawn programs, grants, pilot programs, clearinghouses, awards and contests, and reuse programs)
- Business source reduction programs
- Education (for consumers and in schools)
- Economic incentives and disincentives (variable waste disposal fees, taxes, deposit/refund systems, tax credits, and financial bonuses)
- Regulatory measures (required source reduction plans, labeling, bans, and packaging initiatives)
- Programs aimed specifically at reducing toxics in the waste stream

Most of the project descriptions include information on how to contact the individuals responsible for the programs to obtain more information.

Part III, a "Source Reduction Planning Checklist," presents, in a convenient summary format, an outline of the essential components of effective source reduction programs and a list of specific strategies that can be selected in accordance with local needs. It can thus serve as a tool to help solid waste planners who would like to identify as many source reduction options as possible for their programs — they can literally "check off" program areas and specific initiatives as they include them in their plans. Similarly, the checklist can serve as a reference for officials charged with evaluating and approving plans prepared by others, and for environmentalists and citizens' groups seeking a way to evaluate programs in their communities.

Finally, an appendix contains a bibliography that lists books, reports, and articles that may be helpful to people developing source reduction plans.

Notes

1. US Environmental Protection Agency, *Characterization of Municipal Solid Waste in the United States: 1990 Update,* Washington DC, June 1990, p. 79.

2. US Environmental Protection Agency, *The Solid Waste Dilemma: An Agenda for Action,* Washington DC, February 1989, pp. 16-19.

3. Alabama, California, Colorado, Connecticut, Delaware, Florida, Illinois, Iowa, Kentucky, Maine, Massachusetts, Michigan, Minnesota, Nebraska, New Jersey, New York, North Carolina, Ohio, Oregon, Pennsylvania, Rhode Island, Tennessee, Vermont, Virginia, Washington, Wisconsin, and Wyoming.

■ Chapter 2 The US Solid Waste Problem and Source Reduction

With a solid waste problem of ever-increasing proportions facing the United States, an understanding of the sources of the problem — the sectors of the community and the materials and products that contribute to it — is vital for developing effective solutions. Similarly, an understanding of what source reduction is — and is not — is essential for creating and implementing successful source reduction programs.

❖ The US Solid Waste Problem

The United States is one of the most wasteful countries on earth, according to data on municipal solid waste generation per capita gathered by the Organization for Economic Cooperation and Development (OECD) for 1989. **Table 2-1** lists waste generation information for the OECD countries, the United States, Canada, and the European Economic Community (EEC). The comparison between the United States and the 20 other countries is striking. West Germans, for example, generated only 37 percent, and Italians only 35 percent, of the per capita amount generated by United States residents in 1989. Overall, North Americans (residents of Canada and the United States) generated more than twice as much waste per capita as the Europeans in OECD countries.

The OECD has developed methodologies to make data from different countries more comparable, but international comparisons cannot be taken too literally due to discrepancies in definitions and data collection between countries. For example, United States waste generation data from the OECD and other foreign sources differ from the waste generation data published by the US Environmental Protection Agency. (This report always indicates the source of such figures.) There is little question, however, that the United States is an extraordinarily large producer of waste.

Table 2-1 Annual Waste Generation in Selected Countries (1989)

Country	Amount (pounds per capita)
United States	1920
Canada	1389
Finland	1120
Norway	1051
Denmark*	1042
Luxembourg	1036
The Netherlands	1033
Switzerland	942
Japan	876
United Kingdom	793
Austria	789
Turkey	784
Belgium	776
Spain	716
West Germany	707
Sweden*	704
Greece	698
Ireland*	691
France	673
Italy	669
Portugal*	513
North America	1867
OECD	1151
OECD Europe	747
EEC	727

*Data from 1985

OECD, Organization for Economic Cooperation and Development; EEC, European Economic Community.

Source: *Organization for Economic Cooperation and Development,* OECD Environmental Compendium 1991, *Paris, 1991, p. 133.*

Wastefulness is not a necessary result of our high standard of living, as other countries with high gross national product (GNP) per capita generate far less waste. The ratio of per capita GNP to waste generation was 19 percent higher in West Germany and 129 percent higher in Switzerland in 1984 than in the United States.[1] Switzerland, with a per capita GNP roughly equal to that of the United States ($16,000 per person), generated less than half as much waste per capita (741 pounds) as the United States (1641 pounds) in that year. Relationships between income and waste generation are complex, but it cannot be assumed that greater wealth must automatically result in greater wastefulness.

United States Waste Generation Trends

In order to identify possibilities for reducing waste, it is necessary to understand both the growth and composition of the waste stream. Materials and products accounting for the largest percentages of the waste stream, those that are growing most rapidly, and those that may be reduced most easily may be considered prime targets. Data on waste generation and composition in the United States are published by the US Environmental Protection Agency based on a model developed by Franklin Associates Ltd.[2] The following discussion uses the EPA data, even though the materials flow methodology on which they are based is problematic,[3] because they provide the only comprehensive information available.

Table 2-2 provides a breakdown of the waste stream by materials, by weight. For 1988, the largest categories were paper (40.0 percent) and yard waste (17.6 percent), which together accounted for almost 60 percent of the total waste. As Table 2-2 indicates, all of the material categories increased from 1960 to 1988, but paper and plastic increased the most in actual tonnage and as a percentage of the waste stream. Paper increased 42 million tons, from 34.1 percent to 40.0 percent of the waste stream; plastics increased 14 million tons, from 0.5 percent to 8.0 percent of the waste stream.

EPA projections indicate all of the materials in Table 2-2 are expected to continue to increase in absolute amount over the next 20 years except for glass. Paper and plastic are also expected to increase as a percentage of the waste stream,

Table 2-2 *Materials in the US Waste Stream, by Weight, 1960-2010*

	1960		1988		1960-1988 Change in Weight		2010 Projection	
	Weight (million tons)	% of waste stream	Weight (million tons)	% of waste stream	Million tons	%	Weight (million tons)	% of waste stream
Materials								
Paper and paper-board	29.9	34.1%	71.8	40.0%	41.9	140%	121.2	48.4%
Glass	6.7	7.6%	12.5	7.0%	5.8	87%	9.5	3.8%
Metals	10.5	12.0%	15.3	8.5%	4.8	46%	17.5	7.0%
Plastic	0.4	0.5%	14.4	8.0%	14.0	3500%	25.7	10.3%
Other materials*	8.1	9.2%	20.8	11.6%	12.7	157%	27.0	10.7%
Other Wastes								
Food	12.2	13.9%	13.2	7.4%	1.0	8%	13.7	5.5%
Yard waste	20.0	22.8%	31.6	17.6%	11.6	58%	36.0	14.4%
Total†	87.8		179.6		91.8	105%	250.6	

* This includes rubber, leather, textiles, wood, and miscellaneous inorganics (stones, pieces of concrete, potting soil, etc.).
† Percents may not add up to 100 due to rounding.

Source: US Environmental Protection Agency, Characterization of Municipal Solid Waste in the United States: 1990 Update, *Washington, DC, June 1990, pp. 10, 59.*

while the other materials decline. In other words, only paper and plastic are expected to grow faster than the waste stream as a whole.

Looking at products rather than materials gives a different perspective on how the United States waste stream is changing **(Table 2-3).** The largest increase between 1960 and 1988, 186 percent, was for nondurables, products lasting less than three years. This category includes such items as newspapers; office paper; clothing; paper towels, plates, and cups; and disposable diapers. Next came durables, which increased 165 percent. These include items such as appliances, furniture, and tires. Containers and packaging rose 108 percent, roughly equivalent to the increase in the total waste stream. Food and yard waste and miscellaneous inorganics combined rose only 42 percent, dropping as a percentage of the waste stream. Short-lived products, such as packaging and nondurables, accounted for about 60 percent of the total waste stream in 1988.

Table 2-3 *Products in the US Waste Stream, by Weight, 1960-2010*

	1960		1988		1960-1988 Change in Weight		2010 Projection	
	Weight (million tons)	% of waste stream	Weight (million tons)	% of waste stream	Million tons	%	Weight (million tons)	% of waste stream
Products								
Durable goods	9.4	10.7%	24.9	13.9%	15.5	165%	35.7	14.3%
Nondurable goods	17.6	20.0%	50.4	28.1%	32.8	186%	86.3	34.4%
Containers and packaging	27.3	31.1%	56.8	31.6%	29.5	108%	75.8	30.2%
Other Wastes								
Food, yard waste, and miscellaneous inorganics	33.5	38.2%	47.5	26.5%	14.0	42%	52.8	21.1%
Total*	87.8		179.6		91.8	105%	250.6	

* Percents may not add up to 100 due to rounding.

Source: US Environmental Protection Agency, Characterization of Municipal Solid Waste in the United States: 1990 Update, *Washington, DC, June 1990, pp. 28, 63.*

Within the three main product categories (durables, nondurables, and containers and packaging), only six subgroups accounted for more than 4 percent each of the waste stream in 1988:

Corrugated boxes	12.9 percent
Newspapers	7.4 percent
Miscellaneous durables	5.9 percent
Furniture	4.2 percent
Office paper	4.1 percent
Beer and soft drink containers	4.1 percent

Of these six largest subgroups, the top three — corrugated boxes, newspapers, and miscellaneous durables (such as small appliances, televisions, and video cassettes) — accounted for 26 percent of the waste in 1988.

The fastest growing of these waste subgroups have been corrugated boxes, office paper, and beer and soft drink containers. (Furniture is also cited in the Franklin Associates analysis for EPA as a high growth category, but INFORM questions the use of the materials flow methodology for furniture.)[4] **Table 2-4** shows these fast-growing sectors as a percentage of the waste stream in 1960 and 1988, with projections for 2010. These three product categories are important both

Table 2-4 *Selected Product Categories in the US Waste Stream, by Weight, 1960-2010*

	1960		1988		1960-1988 Change in Weight		2010 Projection	
	Weight (million tons)	% of waste stream	Weight (million tons)	% of waste stream	Million tons	%	Weight (million tons)	% of waste stream
Corrugated boxes	7.3	8.3%	23.1	12.9%	15.8	216%	39.9	15.9%
Office paper	1.5	1.7%	7.3	4.1%	5.8	387%	16.0	6.4%
Beer and soft drink containers	2.1	2.4%	7.3	4.1%	5.2	248%	5.3	2.1%
Total	10.9	12.4%	37.7	21.1%	26.8	246%	61.2	24.4%

Source: US Environmental Protection Agency, Characterization of Municipal Solid Waste in the United States: 1990 Update, *Washington, DC, June 1990, pp. 43. 66.*

because of their relatively large share of the waste stream and because of their growth over the last 30 years.

The EPA projects a continuation of the growth in corrugated boxes to 2010. This would have a major impact on total waste since corrugated was the largest single product category in the waste stream at 23 million tons (12.9 percent) in 1988 and is projected to increase to 39.9 million tons (15.9 percent) in 2010.

Despite frequent predictions that the electronic revolution in offices that made desktop computers ubiquitous would decrease paper use, it has in fact been accompanied by an explosion in office paper waste. Along with equipment that reduces paper use by communicating and storing information electronically came high-speed copiers, laser printers, and fax machines, all of which vastly increased paper use. Office paper waste rose from only 1.5 million tons in 1960 to 7.3 million tons in 1988, an increase of almost 400 percent. As a percentage of waste, it increased from 1.7 percent of the waste stream in 1960 to 4.1 percent in 1988 and is projected at 6.4 percent (16 million tons) in 2010.

Beer and soft drink containers, which increased from 2.4 percent of the waste stream (2.1 million tons) in 1960 to 4.1 percent of the waste stream (7.3 million tons) in 1988, are expected to decline to 2.1 percent of the waste stream in 2010 (5.3 million tons). This reflects several trends moving in opposite directions and provides an illustration of the complex factors influencing changes in the magnitude and composition of the US waste stream.

Three trends have increased waste in this sector: an increase in consumption of beer and soft drinks, a decrease in refillable bottles, and a decrease in the number of times each bottle is refilled. Refillable containers dropped from 87 percent of soft drink containers to under 10 percent between 1960 and 1988, and from 54

Table 2-5 *Beer and Soft Drink Containers, by Material, 1960-2010*

	1960		1988		2010 Projection	
	Weight (million tons)	% of waste stream	Weight (million tons)	% of waste stream	Weight (million tons)	% of waste stream
Glass	1.4	1.6%	5.4	3.0%	2.0	0.8%
Steel	0.6	0.7%	0.1	0.1%	0.2	0.1%
Aluminum	0.1	0.1%	1.4	0.8%	2.2	0.9%
Plastic	0.0	0.0%	0.4	0.2%	0.9	0.3%
Total	2.1	2.4%	7.3	4.1%	5.3	2.1%

Source: US Environmental Protection Agency, Characterization of Municipal Solid Waste in the United States: 1990 Update, *Washington, DC, June 1990, pp. 42-43.*

percent of beer containers to about 5 percent. The average number of times a container was refilled dropped from about 16 to 8 for soft drinks and from 23 to 6 for beer in that same period.[5]

The increase in beer and soft drink containers from 2.1 million tons in 1960 to 7.3 million tons in 1988 would have been even larger had it not been for two trends that decreased this waste: the use of larger bottles for soft drinks and "lightweighting," the shift from heavier glass and steel to lighter plastic and aluminum containers. The projected decline of beer and soft drink containers to 5.3 million tons in 2010 reflects a continuation of lightweighting and a tapering off of the effect of the decline of refilling. The big increase in beer and soft drink containers occurred between 1960 and 1980. After lightweighting began, there was a 15 percent decrease in these containers between 1980 and 1988, the trend that is expected to continue.

As **Table 2-5** shows, the impact of these conflicting trends has been dramatic. Glass beer and soda bottles rose from 1.4 million tons in 1960 to 5.4 million tons in 1988, but are projected at only 2.0 million tons in 2010 — less than half the current level. At the same time, aluminum containers increased from 0.1 million tons in 1960 to 1.4 million tons in 1988, and are expected to continue to increase to 2.2 million tons in 2010. Plastic containers increased from nothing in 1960 to 0.4 million tons in 1988 and are projected to more than double to 0.9 million tons in 2010.

Another trend worth noting is the switch from reusable to disposable "convenience" products. Miscellaneous nondurables, while not one of the six largest waste product subcategories, have grown rapidly. This EPA product subcategory is composed of products with a life of under 3 years. They are made mainly of

Table 2-6 *Summary of Products Discarded* in Municipal Solid Waste, 1988*

	Percent of Discards by Volume	Percent of Discards by Weight
Nondurable goods	34.0	27.6
Containers and packaging	29.6	27.6
Durable goods	22.2	14.7
Yard wastes	10.4	20.0
Food wastes	3.3	8.5
Other	0.5	1.7

* Discards after recycling and composting, before combustion and landfilling.

Source: US Environmental Protection Agency, Characterization of Municipal Solid Waste in the United States: 1990 Update, *Washington, DC, June 1990, p. 89.*

plastic, and include such items as pens, razors, lighters, disposable medical supplies, and novelty items.

Miscellaneous nondurables surged upward from only 0.5 percent of waste in 1960 to 2.5 percent in 1988. Their growth is expected to continue to a projected 2.9 percent in 2010. For example, United States residents discard billions of razors and pens each year. Disposable diapers, not generally available in 1960, rose to 2.7 million tons (1.5 percent of the waste stream) in 1988, but are expected to decline to 2.4 million tons by 2010 due to lighter weight diapers and a decline in the birth rate.[6]

Despite growing concerns about solid waste, new products with very short lives are still replacing older, more durable products. Cameras, which just a few years ago were designed to last a lifetime, now are available in disposable versions designed to take only one roll of film. Five million single-use cameras were sold in 1989, just 2 years after they were introduced. Their use tripled by 1991, when United States consumers bought 15 million of these cameras.[7]

Weight versus Volume

Volume is a more relevant measure than weight for solid waste planners concerned about trucking capacity and landfills: landfills fill up rather than get too heavy. However, the preceding discussion of waste generation and composition was based on weight measurements because there are no waste generation and composition data over time measured in volume. Volume measurements are more difficult to make than weight measurements since volume is dependent on varying

Table 2-7 *Summary of Materials Discarded* in Municipal Solid Waste, 1988*

	Percent of Discards by Volume	Percent of Discards by Weight	Volume: Weight Ratio
Paper and paper-board	34.1	34.2	1.0
Plastics	19.9	9.2	2.2
Yard wastes	10.4	19.9	0.5
Ferrous metals	9.8	7.0	1.4
Rubber and leather	6.4	2.9	2.3
Textiles	5.3	2.5	2.1
Wood	4.1	4.2	1.0
Food wastes	3.3	8.5	0.4
Aluminum	2.3	1.1	2.1
Glass	2.0	7.1	0.3
Other	2.5	3.6	0.7

* This assumes that all waste (after recycling) is landfilled.

Source: US Environmental Protection Agency, Characterization of Municipal Solid Waste in the United States: 1990 Update, *Washington, DC, June 1990, p. 90.*

compaction rates. Measurement standards for weight have been established (pounds and tons), while those for waste volume have not.

The relationship of weight and volume of materials and products discarded varies widely. **Tables 2-6 and 2-7** illustrate this variation using data from studies of materials and products in landfills in 1988. For example, food and yard wastes represent 28.5 percent of the waste stream by weight, but only 13.7 percent by volume. Plastics, which are only 9.2 percent by weight, are 19.9 percent by volume.

The weight versus volume issue has particular relevance to source reduction. Large weight reductions have been achieved, particularly in lightweighting packages, but this has often been accompanied by even greater increases in volume. This is primarily due to shifts from heavy materials such as glass and metals to lighter-weight plastics. However, bulky, light-weight plastic containers and foam products are costly to collect and take up large amounts of space in collection trucks and landfills, so the weight reductions achieved do not always alleviate the solid waste problem.

If source reduction is to help solve the solid waste problem, the implications for volume will have to be considered as well as those for weight. An ideal source reduction strategy would decrease weight without increasing volume or decrease volume without increasing weight. An example of a source reduction strategy that meets this standard is the elimination of nonessential layers of packaging.

Figure 2-1 *Municipal Solid Waste Management, 1960 to 1995, in millions of tons*

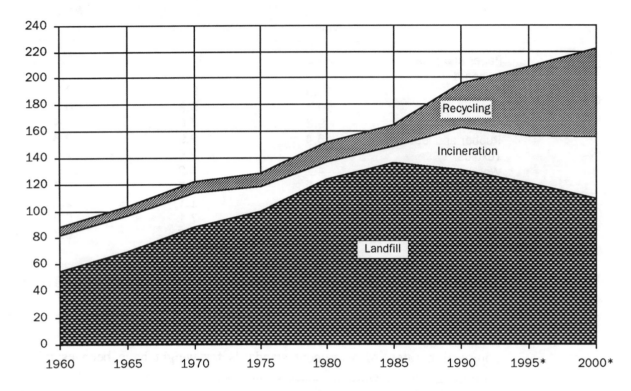

* Estimated

Source: Franklin Associates, Ltd., Characterization of Municipal Solid Waste in the United States, 1992 Update, *p. 3-2*

Declining Disposal Capacity

While the municipal solid waste stream has more than doubled since 1960, landfills, the primary disposal facilities, have been closing. Only 6600 were in operation in 1989, down from 20,000 a decade earlier. As greater concern over environmental impacts has led to higher costs and fewer available disposal sites, the days of just dumping garbage in convenient open spaces have come to an end.

In the 1970s, interest developed in waste-to-energy plants (incinerators generating energy by burning garbage) as the primary response to closing landfills. At the end of 1990, there were 128 waste-to-energy plants operating and approximately 100 plants on the drawing boards[8]; the EPA projected incineration would almost double between 1988 and 1995, from 25.5 to 45.5 million tons per year. Some new incinerators are being constructed, but public opposition to both landfills and incinerators remains intense. Between 1987 and 1991, plans for 121 proposed waste-to-energy facilities were canceled.[9]

Figure 2-1 illustrates municipal solid waste management from 1960 to 1990, with projections to 1995. It indicates that the United States burns slightly less today than it did in 1960, although the waste stream has doubled. The amount landfilled has declined from the peak of the mid-1980s, but still accounts for 73 percent of the total waste.

If the upward trend of waste generation continues along with the downward trend in landfilling and the resistance to incineration, can the gap between waste generated and waste landfilled be closed by recycling? If not, a major disposal crisis looms.

Recently, recycling has been regarded by many as a panacea: "Recycling has moved into the same league as motherhood, the flag and apple pie," claimed a 1991 article in *The New York Times*.[10] While recycling is a vital strategy for conserving energy and natural resources, and can clearly help to reduce the need for disposal capacity, it has proven costly and may be polluting. New York City, with some of the highest recycling costs in the nation, has estimated its recycling costs at about $300 a ton.[11] This should decrease as volume increases; however, recycling may remain relatively expensive for the foreseeable future. Further, INFORM research indicates that current recycling rates are sometimes overstated and recycling costs per ton understated when they are based on materials collected for recycling, not materials actually sent to market and remanufactured into new products.[12]

A recent study by the Regional Plan Association, on the other hand, estimates that, over the next 25 years, recycling in the New York-New Jersey-Connecticut region will not be more expensive than incineration or landfilling;[13] however, this is an area with high disposal costs. (The RPA study also estimates large savings from source reduction and is discussed in more detail in the following section on "The Source Reduction Option.")

Further, recycling requires its own facilities which have their own environmental impacts: materials recovery facilities, municipal composting sites, and de-inking plants, among them. Communities have also expressed concern about siting some of these; for example, Islip, New York, residents have complained about odors from a municipal composting facility.[14]

Increasing Waste Management Costs

Growing waste, decreasing disposal capacity, and more stringent environmental regulations have led to dramatic increases in waste management costs along with increasing regional disparities. **Table 2-8** illustrates the rise in landfill tipping fees by region from 1986 to 1990. The increases in just these 4 years ranged from 43 percent in the South to 269 percent in the Northeast. The regional gap has, in fact, widened since 1986, when the highest tip fee was 3.5 times the lowest. By 1990, the highest was almost 6 times the lowest due to the huge tip fee increase in the Northeast. Nationwide, average per ton waste management costs (including

Table 2-8 *Average Landfill Tipping Fees in the United States, 1986-1990 ($/ton)*

Region	1986	1988	1990	% Increase 1986-1990
Northeast	$17.57	$61.11	$64.79	269%
Mid-Atlantic	$21.41	$33.84	$40.75	90%
South	$11.86	$16.46	$16.92	43%
Midwest	$11.75	$15.80	$23.02	96%
West Central	$6.21	$10.63	$11.06	78%
South Central	$8.71	$11.28	$12.50	44%
West	$11.10	$19.45	$30.63	176%

Source: "National Solid Waste Management Association (NSWMA) Survey Tracks Upward Trend in Landfill Tipping Fees," Integrated Waste Management, *December 11, 1991, p. 7.*

collection, recycling, and/or disposal) have been estimated at $46 to $107 in 1986 and at $59 to $194 in 1991; these costs are projected to rise further to $128 to $219 by 1996.[15]

❖ The Source Reduction Option

The increasing amounts and toxicity of municipal solid waste, increasing disposal costs, and increasing opposition to siting waste management facilities all point to source reduction — reducing the waste generated — as a solution to the waste problem that deserves serious consideration. Yet surveys indicate that only about one in five people in the United States has any idea of what source reduction means.[16] So source reduction remains the universally acclaimed priority, infrequently implemented in practice.

The savings from source reduction can be substantial. New York City's 20-year solid waste plan estimates that 7 percent source reduction would save the city's waste management system approximately $90 million in avoided waste management costs in the year 2000 alone. The cumulative avoided costs between 1992 and 2010 would be $700 to $800 million (in net present value terms).[17] These savings are based on a relatively low source reduction figure of 7 percent. If more aggressive source reduction is accomplished, such as 15 to 20 percent, savings could be in the billions of dollars.

The Regional Plan Association, an independent research organization covering the tristate area that includes parts of New York, New Jersey, and Connecticut,[18] has published an analysis of solid waste management in that region. It estimates that if the three states meet their individual, legislatively mandated source reduction goals (which range from 10 to 26 percent of projected waste), annual savings will be $190 million by 2000 and total savings between 2000 and

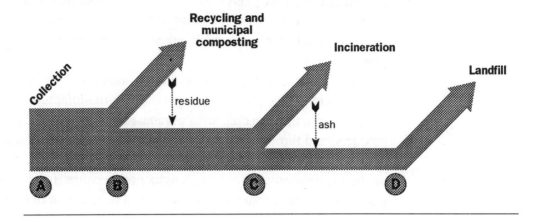

2015 will reach \$4.24 billion.[19] All of these savings relate only to avoided waste management costs — savings on collection and disposal. Savings in purchasing costs, which could be much greater, are not included.

What Source Reduction Is

Much confusion exists as to what source reduction is and if and how it differs from "waste reduction" and "recycling." In this study, INFORM uses the term "source reduction" to mean "a reduction in the amount and/or toxicity of waste entering the waste stream—or waste prevention" (this is consistent with the definition used by the EPA).

Figure 2-2 presents a simplified representation of the municipal waste stream. Waste is generated at "the source" (A), the point where it enters a waste collection system. Parts of the waste stream are diverted to composting and recycling (B) and incineration (C), with the remainder flowing to point (D) where it is sent to a landfill. Residues from both recycling and incineration reenter the waste stream. Only initiatives that reduce the amount of waste at point (A) are source reduction; anything that happens to waste beyond point (A) is not source reduction because the waste has already been generated and must be collected and processed.

There are, however, some gray areas in defining source reduction, which generally relate to use. Activities that require some collection outside the public collection system and that lead to reuse without manufacturing can be considered source reduction. For example, reuse of clothes and equipment through donations to charitable organizations or swaps are source reduction since no remanufacturing is required, as is returning beverage bottles for refilling (but not recycling).

In defining source reduction, it is important to understand its dual meaning: it refers not only to the amount of waste generated, but also to its toxicity. Sometimes reductions in amount may be accompanied by increases in toxicity. In lightweighted packages, for example, the lighter-weight plastics may contain more toxic constituents than the glass they replace. Such potential tradeoffs between reductions in amount and toxicity cannot be quantified, but preferable source reduction strategies would not accomplish either reduced amount or toxicity at the expense of the other.

Reductions at the source in the amount and/or toxicity of municipal solid waste can be accomplished in many ways: through manufacturer redesign of products and packages; through consumer purchases of less wasteful products and reuse of products; and through institutional changes in practices (such as using paper on both sides) and purchases of more durable and less toxic products. Extending product life accomplishes source reduction because, for example, a refrigerator that lasts 20 years creates half the waste of a refrigerator of the same size that lasts 10 years. Product life can be extended through design and manufacturing processes as well as through better maintenance and repair.

In setting source reduction strategies, the first priority is elimination; the second, reuse or increased intensity of use. Office paper provides a good example. Reducing the number of copies of documents by establishing central filing systems or storing data electronically saves twice as much paper per copy as copying the same number of documents on two sides. That is, not making one copy of a two-page document saves two sheets of paper, while copying it two-sided saves only one sheet because it uses one sheet of paper. Both accomplish source reduction, but they achieve different amounts.

What Source Reduction Is Not

To implement effective policy, determining what source reduction is not is as important as determining what it is. It is not recycling or buying products with recycled content. It is not municipal composting. It is not household hazardous waste collection. It is not beverage container deposit and return systems, as accomplished by currently operating "bottle bills" in the United States.

While "source reduction" is a term referring specifically to the generation of waste, "waste reduction" is an all-encompassing term that can be measured at different points on the waste stream continuum shown in Figure 2-2. Using the terms "source reduction" and "waste reduction" interchangeably, as is frequently done, causes confusion as to what source reduction really means and fails to properly distinguish it from recycling.

Nevertheless, some states and communities define the goals of their waste programs as "waste reduction." Their concern is with reducing the amount of waste entering some or all of their waste disposal facilities. Ohio, California, and North Dakota, for example, have 25 percent waste reduction goals measured by waste entering landfills (point D in Figure 2-2). In these states, waste reduction includes recycling, composting, and incineration as well as source reduction. New

York State has a goal of 50 percent waste reduction measured by the amount of waste entering either landfills or incinerators (point C in Figure 2-2). New York, therefore, defines waste reduction as including recycling, composting, and source reduction. New York also has a separate source reduction goal of 8-10 percent (to be measured at point A).

Recycling is a strategy for managing materials that would otherwise be treated as waste, but it does not reduce the amount of waste generated. It reduces the amount of materials requiring disposal and is a vital strategy for conserving energy and natural resources. But recyclables must be collected, processed, and remanufactured into new materials or products. These processes incur costs and may, themselves, cause pollution.

Packaging provides an example of how the implications of source reduction policy differ from those of recycling. Manufacturers often claim that a package is "environmentally friendly" because it is made of recycled content or is recyclable. However, recycling cannot justify overpackaging (such as the use of extra unneeded layers of packaging) if the hierarchy that puts source reduction before recycling is taken seriously. Following the hierarchy would require that packaging be reduced as much as possible before moving to a recycling strategy.

For example, consider a product such as a shampoo or a bath gel that is sold in a bottle that is wrapped in corrugated and then inserted in a box that, in turn, is shrink-wrapped. A source reduction strategy would aim to eliminate the three excess layers of packaging — the box, corrugated, and shrink wrap — leaving the product in a single container. A recycling strategy alone would require that all the layers be recyclable or made of recycled materials, but would not eliminate them.

Composting also illustrates the distinctions between source reduction and recycling. Municipal composting is a form of recycling. It requires the collection and processing of grass clippings and leaves, and sometimes marketing as well. Backyard composting, however, is source reduction. When individuals keep grass clippings and leaves in their own backyards and then reuse the decomposed material, nothing has entered the recycling system or waste stream. The municipality has not collected, processed, or disposed of the material, so source reduction has been achieved.

Separate collection of household hazardous wastes is often cited as a source reduction initiative. While this activity is commendable, it is not source reduction. Household hazardous waste collections reduce the toxicity of waste entering disposal facilities and, thereby, reduce the pollutants generated by incinerators and landfills. However, the hazardous waste must still be collected and disposed of — often in special hazardous waste facilities. Thus, the materials are removed from the solid waste stream but put into the hazardous waste stream. True hazardous waste source reduction would substitute nontoxic materials for toxics in the manufacturing process or reuse materials, such as old paint, so that they do not enter the waste stream at all. Using a mechanical pencil sharpener that does not require batteries is source reduction. Bringing batteries to a household hazardous waste collection site is not, since the batteries require management and disposal.

"Bottle bills" requiring deposits on beverage containers are frequently described as source reduction initiatives. They are not necessarily source reduction, however, since most of the bottles and cans presently collected are recycled, not refilled. This is one of the gray areas discussed above, since the bottles are collected outside the public collection system. However, the bottles are being remanufactured into new products, rather than reused. If the containers were washed and refilled, bottle bills could accomplish source reduction.

Source Reduction and Lifecycle Analysis

Major research efforts by government and the private sector are now underway to develop methodologies that could determine the environmental impacts of a product or package over its entire life. This would include analysis of the raw materials, energy, and water used; the environmental, worker, and public health impacts of production; and the environmental and public health impacts of solid waste management.

At present, lifecycle analysis is costly, controversial, and cumbersome. Study of a single product may cost $25,000 to $100,000[20] and there may never be agreement on the relative significance of different pollutants or on the scope of the analysis. For example, how do we compare air pollution from trucks to amounts of solid waste generated? If the loss of trees counts in evaluating lifecycle impacts of paper, do the environmental impacts of manufacturing the equipment used to cut the trees also count? How do we quantify the effects of depleting rainforests? How do we rank the toxicity of pollutants when data are often unavailable and, at best, highly uncertain and variable? Far from being a science, lifecycle analysis requires difficult judgments that leave the system open to political and financial pressures.

Tellus Institute, a nonprofit research and consulting organization, published in June 1992 a lifecycle analysis that evaluated the relative importance of different environmental impacts of packaging. This required establishing toxicity factors and remediation costs for each of the individual pollutants in packaging materials. The study found that costs per pound of pollutants varied from less than $1.00 to over $200,000.[21] These cost estimates are also highly changeable; for example, the Tellus study originally estimated the costs of lead at $25,000 per pound, but Tellus subsequently revised this estimate to $1600.[22] Tellus accomplished a Herculean task in its packaging study, and explicitly stated its assumptions. However, comparative evaluations of different environmental impacts and different pollutants are a matter of judgment, and such judgments are inherently open to interpretation and, hence, are controversial. Thus, developing a generally accepted methodology to quantify the environmental impacts of packaging materials remains an unsolved problem.

Fortunately, there is no need to hold off on efforts to reduce the amount of garbage we generate until there is agreement on acceptable lifecycle methodology. Moving forward on source reduction is not contingent upon lifecycle analyses. Many source reduction strategies are clearly beneficial to the environment. For example, a ceramic mug that is used for several years has less environmental impact and is less expensive than the hundreds or thousands of disposable cups it replaces, even though it must be washed. Eliminating unnecessary layers of packaging reduces waste as well as the consumption of resources. There is virtually no negative environmental impact from using paper on two sides rather than on one.

Lifecycle analysis may be useful when the two source reduction goals of reducing the amount and toxicity of waste come in conflict, since it is preferable that neither of these should be accomplished at the expense of the other. This problem can arise, for example, in the "lightweighting" of packages and containers when a heavy material like glass is replaced by a lighter plastic that may contain more toxic substances. Further work is needed to better evaluate such tradeoffs.

Reducing Toxic Materials in Solid Waste

When household products containing toxic chemicals — such as batteries, paints, cleaners, and home pesticides — are disposed of, the toxic chemicals they contain enter landfills, garbage incinerators, home septic systems, or public sewer systems. From there, they can move into air, land, and water, and may affect public health and the environment. While the actual health and environmental risks associated with this pollution are still in many cases a matter of scientific debate, the EPA and others have measured the concentrations of toxic pollutants discharged by waste disposal facilities. Chapter 13 provides more information on toxics in the municipal solid waste stream and discusses the associated pollution in more detail.

Keeping toxics out of the municipal solid waste stream is thus key to reducing the pollution caused by waste disposal. INFORM research on state-of-the-art garbage burning stressed that, for those communities that choose this disposal strategy, the vital first step in attaining the cleanest possible incineration is to keep materials containing toxic substances or pollutant precursors out of the incinerator waste stream.[23] Similarly, keeping such materials out of landfills prevents leaching into groundwater and other problems.

The primary strategy for accomplishing this is source reduction. Recycling is generally not a viable alternative for toxic substances. In fact, source reduction of toxics can make all forms of waste management safer for public health and the environment and may, thereby, make it easier to site facilities for recycling, incineration, and landfilling materials that cannot be reduced at the source. Strategies for reducing toxic chemicals in the municipal solid waste stream are also discussed in Chapter 13.

Notes

1. Environmental Defense Fund (Richard A. Denison and John Ruston), *Recycling and Incineration: Evaluating the Choices,* Island Press, Washington, DC, 1990, p. 35.

2. For this study, INFORM used 1988 data published in 1990: US Environmental Protection Agency, *Characterization of Municipal Solid Waste in the United States: 1990 Update,* Washington, DC, June 1990, p. 1. However, in July 1992, as this report went to press, Franklin Associates published 1990 data in *Characterization of Municipal solid Waste in the United States: 1992 Update.* Major changes between the 1988 and 1990 data are described here.

   For materials, two components of the "other" category grew significantly as a percentage of the waste stream: wood, from 3.6 to 6.3 percent, and textiles, from 2.1 to 2.9 percent. Franklin attributes these changes to underestimates in previous years. Paper declined from 40 to 37.5 percent of the waste stream, although it increased in actual tonnage.

   For product subcategories, interesting changes took place for corrugated boxes, office paper, and beer and soft drink containers. Corrugated boxes, the largest waste stream component, decreased from 12.9 percent of the waste stream in 1988 to 12.2 percent in 1990, and Franklin now projects that corrugated will not grow as fast as previously estimated, remaining at 12.2 percent of the waste stream in 2000. Office paper decreased from 4.1 to 3.3 percent of the waste stream (and from 7.3 to 6.4 million tons), but this change was due to a reclassification of the components previously included in the office paper category. Finally, while beer and soft drink containers remained virtually constant as a percentage of the waste stream (from 4.1 percent in 1988 to 4.0 percent in 1990), they increased in tonnage for the first time since 1980 (from 7.3 to 7.8 million tons). Further, Franklin now projects a much slower decrease in these containers as a percentage of the waste stream, down to 3.8 percent in 2000.

3. The methodology bases waste generation on production data. In the case of disposables or short-lived products, production is a good proxy for waste, but for some durables, such as furniture (see note 4), it is not.

4. The Franklin analysis is based on the assumption that the average lifetime for furniture is 15-20 years, after which all furniture is thrown away. An increase in furniture production 20 years ago, according to Franklin, is responsible for the spurt in the waste stream now. If the durability of furniture has changed over the past 20 years, this would not be reflected in the Franklin data, nor would the data reflect the reuse of furniture through charitable donations or second-hand shops.

5. INFORM estimates based on studies by OECD and Franklin Associates.

6. US Environmental Protection Agency, *Characterization of Municipal Solid Waste in the United States: 1990 Update,* Washington, DC, June 1990, pp. 36, 64, 65.

7. Joan E. Rigdon, "For Cardboard Cameras, Sales Picture Enlarges and Seems Brighter than Ever," *The Wall Street Journal,* February 11, 1992, p. B1.

8. Jonathan V.L. Kiser, "A Comprehensive Report on the Status of Municipal Waste Combustion," *Waste Age,* November 1990, p. 107.

9. Shearson Lehman Brothers, *The Waste Watcher's Guide to the Waste-to-Energy Market,* September 13, 1991, p. 3.

10. John Holusha, "Mixed Benefits From Recycling," *The New York Times,* July 26, 1991, p. D2.

11. Michael Specter, "Dinkins's Role in Sanitation Is Faulted," *The New York Times,* January 18, 1992, pp. 25-26.

12. INFORM (Maarten de Kadt, Ph.D.), *Recycling Programs in Islip, New York, and Somerset County, New Jersey,* New York, 1991.

13. Regional Plan Association, *Existing and Future Solid Waste Management Systems in the RPA Region,* prepared by Tellus Institute, March 1992, pp. 6-8.

14. INFORM (Maarten de Kadt, Ph.D.), *op. cit.*

15. "National Solid Waste Management Association (NSWMA) Survey Tracks Upward Trend in Landfill Tipping Fees," *Integrated Waste Management,* December 11, 1991, p. 8.

16. Karen S. Brattesani, Ph.D., "Seattle Solid Waste Utility 1990 Waste Reduction Survey," *Research Innovations*, Seattle, April 1990, p. 2. Frank Edward Allen, "Environmental Terms Catch on Very Slowly," *The Wall Street Journal*, July 11, 1991, p. B1.

17. New York City Department of Sanitation, *A Comprehensive Solid Waste Management Plan for New York City and Final Generic Environmental Impact Statement*, August 1992, p. 17.2-2.

18. The RPA area is a 31-county, 20 million person region comprising western Connecticut, Long Island, New York City, southeastern New York, and the northern half of New Jersey.

19. Regional Plan Association, *op. cit.*, p. 12.

20. "The Green Cross Certification Company (Oakland, Calif.) Will Offer a Product Eco-Label in the U.S. Based on the Life Cycle Analysis," *Crosslands European Environmental Bulletin*, Sept. 12, 1991, p. 2.

21. Tellus Institute, *CSG/Tellus Packaging Study*, prepared for the Council of State Governments, the US Environmental Protection Agency, and the New Jersey Department of Environmental Protection and Energy, Boston, MA, 1992.

22. Personal communication. John Schall, Tellus Institute, with Bette K. Fishbein, INFORM.

23. INFORM (Marjorie J. Clarke, Maarten de Kadt, Ph.D., and David Saphire), *Burning Garbage in the US: Practice vs. State of the Art*, New York, 1991, pp. 42-48.

▨ Chapter 3 Policy, Goals, and Measurement

Planning is central to developing effective source reduction programs. Before source reduction planners start developing specific source reduction initiatives for their communities, it is extremely important that they know what they are trying to reduce, how much reduction they want to achieve, and how they will measure their results. While most municipal solid waste plans endorse the EPA hierarchy that makes source reduction the first priority, many never get beyond this endorsement. They need an explicitly stated source reduction policy, clearly defined goals, and meaningful measurement strategies. Without these, they will have difficulty evaluating the effectiveness of their programs.

❖ Source Reduction Policy

The first step in planning for source reduction is a clear statement of policy, including a definition of terms that clarifies what source reduction means so that it can be differentiated from other waste management options, such as recycling, as described in Chapter 2. In other words, instead of a policy of "diversion from landfills," which leaves ambiguity as to whether the strategy should be reduction, recycling, or incineration, a clear policy would state explicitly that its aim is source reduction, include a definition of that term, and then specify goals and measurement methodology.

❖ Setting Source Reduction Goals

The second step in source reduction planning is establishing goals. Clearly defined goals include a percent reduction, a baseline year, a target year, a method of measurement, and specified waste streams. It is also vital that source reduction goals be separate from recycling or "waste reduction" goals; if source reduction is truly the top priority, it requires its own goal. When there is a combined goal for source reduction and recycling, source reduction is often neglected in terms of staff and budget. (This discussion of goals and measurements refers to reduction in the amount of waste; reductions in toxicity are discussed in Chapter 13.)

Components of Goals

Baseline years and target years are essential for a meaningful source reduction goal. By itself, a goal of 25 percent source reduction means very little: 25 percent by when? from when? of what? Stating the measurement parameters is also necessary to define a goal: how is this 25 percent reduction measured? It could be a reduction from current total waste stream levels, a reduction in per capita waste generation, or a reduction from a projected increase. Measurement issues are discussed in more detail in the following section, starting on page 31.

Defining the applicable waste streams to be reduced is also important: who produces waste and what materials are involved? Should separate goals be set for different generating sectors: residential, commercial, institutional? Should separate goals be set for specific materials like paper, glass, metals, plastic, and organics?

The Importance of Waste Composition

Knowledge of the composition of the waste stream is very helpful in setting realistic goals because it allows communities to set source reduction priorities. Materials can be targeted for source reduction because they constitute a major proportion of the waste stream, are easy to reduce, or are major contributors to pollution during disposal. Since the waste stream varies from community to community, in-depth information about it requires a waste audit — an actual sampling of waste generated to determine its composition by material, product, and generating sector.

Yard waste, for example, is a good target for source reduction. When burned, it creates emissions of nitrogen oxides (NO_x); it represents a large component of the waste stream (almost 18 percent nationally); and it can quite readily be reduced through backyard composting. A suburban community with a large proportion of yard waste can set a higher overall source reduction goal than a densely populated city with a small proportion of yard waste.

Goals in the US Today

While 38 states and the District of Columbia have "waste reduction" goals of some sort, only seven have specific, separate source reduction goals: Connecticut, Maine, Massachusetts, Michigan, New Jersey, New York, and Pennsylvania. These states all identify target years; however, they do not all clearly specify the baseline year or state how source reduction is to be measured. (These specific source reduction goals refer only to reduction in the amount of waste, not to toxicity reduction.) Rhode Island and Wisconsin are also actively pursuing source reduction, although they deliberately have not set separate goals. In comparison, 23 states and the District of Columbia have recycling goals, and 17 states have waste reduction goals; these do not differentiate between the amount designated for recycling and that designated for source reduction.

Table 3-1 shows source reduction, recycling, and waste reduction goals by state. Waste reduction has different meanings in different states: it usually includes source reduction and recycling. In three states (Ohio, North Dakota, and California), it sometimes includes incineration. New technologies, such as mixed waste composting, may or may not be included in waste reduction. The table also shows the 38 states that require or give incentives to their localities to develop solid waste management plans. The goals that states set provide the guidelines for their localities. For example, in New York, which has a specific 8-10 percent source reduction goal, localities usually include the state goal in their plans.

Table 3-1 Waste Reduction Goals in the United States

| State | Goals | | | Local Plans Required |
	Source reduction	Recycling	Waste reduction	
Alabama		25% by 1995		■
Alaska				
Arizona				
Arkansas		30% by 1995 40% by 2000		■
California			25% by 1995* 50% by 2000 from 1990 baseline of total waste	■
Colorado				
Connecticut	No net increase in per capita waste, 1990-2010	37% by 2010		■
Delaware		30% by 1994†		†
Dist. of Columbia		45% by 1994		†
Florida		30% by 1994		■

continued

* Diversion from landfill, 10% allowed through incineration although the California Integrated Waste Management Board says any increase in incineration is unlikely.
† This is not a legislated goal, but a goal that the Solid Waste Authority, a quasi-government agency, hopes will be reached.
‡ Solid waste planning is being done on a state (rather than a local) level for the District of Columbia, Delaware, and Rhode Island because they are so small.
** Counties with populations under 100,000 do not have to reach the 25% goal until 2000.
†† Diversion from landfill, although the state has no solid waste incinerators and no proposals to build any.
‡‡ Doesn't include yard waste, but does include incineration.

Table 3-1 *Waste Reduction Goals in the United States (continued)*

| State | Goals | | | Local Plans Required |
	Source reduction	Recycling	Waste reduction	
Georgia			25% by 1996 from 1992 per capita	■
Hawaii				
Idaho				
Illinois			15% by 1994 25% by 1996** from 1991 baseline of total waste	■
Indiana			35% by 1996 50% by 2000, from 1991 baseline of total waste	■
Iowa			25% by 1994 50% by 2000 from 1988 baseline of total waste	■
Kansas				■
Kentucky			25% by 2000 from 1993 baseline of per capita waste	■
Louisiana		25% by 1992		■
Maine	5% by 1992 10% by 1994 from 1990 baseline of total waste	25% by 1992 50% by 1994		■
Maryland		15-20% by 1992 20% by 1994 for state agencies		■
Massachusetts	10% by 2000 from 1990 baseline of total waste	46% by 2000		■
Michigan	5-12% by 2005 from 1989 baseline of total waste		50% by 2005	■

continued

Table 3-1 *Waste Reduction Goals in the United States (continued)*

| State | Goals | | | Local Plans Required |
	Source reduction	Recycling	Waste reduction	
Minnesota		25% by 1994 for greater MN, 35% for Twin Cities, 40% for state agencies		■
Mississippi		25% by 1996		■
Missouri		40% by 1998		■
Montana				
Nebraska				■
Nevada		25% by 1994		■
New Hampshire			40% by 2000 from 1990 baseline per capita	■
New Jersey	Cap waste generation at 1990 baseline by 1996, reduce by 2000	60% by 1995		■
New Mexico			25% by 1995 50% by 2000, baseline of 4 lb/ day per capita	
New York	8-10% by 1997 from 1987 per capita baseline	40-42% by 1997		■
North Carolina			25% by 1993 40% by 2000 from 1991 per capita baseline	■
North Dakota			10% by 1995 25% by 1997 40% by 2000†† baseline of 1991 total waste	■
Ohio			25% by 1995†† from 1989 baseline of total waste	■

continued

Table 3-1 *Waste Reduction Goals in the United States (continued)*

State	Goals			Local Plans Required
	Source reduction	Recycling	Waste reduction	
Oklahoma				■
Oregon		50% by 2000 56% by 2006		■
Pennsylvania	No increase in generation from 1988 to 1997	25% by 1997		■
Rhode Island		15% by 1994 for residential		‡
South Carolina		25% by 1997	30% by 1997 from 1993 baseline of total waste	■
South Dakota			20% by 1995 35% by 2000 50% by 2005 from 1990 baseline of total waste	
Tennessee			25% by 1995 from 1989 base line per capita	■
Texas		40% by 1994		■
Utah				■
Vermont			40% by 2000 from 1987 baseline per capita	■
Virginia		10% by 1992 15% by 1993 25% by 1995		■
Washington			50% by 1995 from 1990 baseline of total waste	■
West Virginia		30% by 2000		■
Wisconsin				■
Wyoming				

Table 3-2 *New York City's Estimates of Source Reduction Potential (by Weight), by Material and Generating Sector, 2000*

Material	Total Waste, 2000* (thousand tons)	Source Reduction Potential (thousand tons)			% Reduction	
		Residential sector	Institutional sector	Commercial sector	Waste material	Waste stream
Paper	3700	100	79	304	13%	6%
Plastics	710	(6)	(5)	(15)	(4%)	(0.3%)
Organics	2700	85	13	36	5%	2%
Glass	320	21	2	4	9%	0.3%
Aluminum	70	0.1	0.1	0.6	1%	0.01%
Other metal	280	5	0.9	4	4%	0.1%
Inorganics	80	2	0.3	0.05	3%	0.03%
Hazardous	20	0.6	0.09	0.2	4%	0.01%
Bulk	370	39	1	0	11%	0.5%
Total Waste*	8300	3400	930	3900		
Total Prevented	670	250	90	330		
Percent Prevented	8%	7%	10%	9%		

* Total waste refers to estimates of waste generation in 2000 if there were no source reduction.

Source: New York City Department of Sanitation, A Comprehensive Solid Waste Management Plan for New York City and Final Generic Environmental Impact Statement, *August 1992, p. 7-11.*

New York City's proposed 20-year solid waste management plan exemplifies a more precise way in which communities can set targeted goals; **Table 3-2** shows the city's estimates of the potential for reduction by material and generating sector. (These estimates come from the version of the city's 20-year plan that was approved by the New York City Council in August 1992; as this report went into production, the plan had not yet been approved by the state Department of Environmental Conservation.)

Numerical source reduction goals can be misleading, however, if terms are not specifically defined. For example, New York State has a source reduction goal of 8 to 10 percent by 1997 from the 1988 baseline. The New York City goal, as illustrated in Table 3-2, is an 8 percent reduction in 2000, calculated as a reduction in the total waste projected for that year without source reduction. Since New York City has projected an 8 percent increase in waste by 2000 from the 1990 baseline, its 8 percent source reduction goal actually represents "no net increase from 1990," not an actual decrease. Thus, although both the state and the city specify goals of 8 percent reductions, their real goals are very different, with the city's being much lower than the state's.

❖ Measuring Source Reduction

Measurement systems are important for effective source reduction programs not only because they help communities set realistic goals for their programs and establish program priorities, but also because they allow communities to track and evaluate the progress of their source reduction activities: to recognize their accomplishments and target areas for further efforts. Yet, measuring source reduction can seem complex, and difficulties in establishing measurement systems have been a major factor hindering widespread source reduction efforts. Communities may be unwilling to commit resources to source reduction programs if they lack a method for measuring and evaluating results.

Source reduction has been effectively measured at the micro level of individual materials or small groups: companies, institutions, or several dozen households. Measurement on a macro level — a community-wide, multimaterial basis — has proven more difficult.

Communities can move ahead with setting up information and measurement systems (described below) that will improve their ability to set source reduction program goals and analyze the effectiveness of their actions. In fact, it is useful for communities to determine their measurement strategy as they design source reduction programs so that measurement is an integral part of their efforts, rather than an issue to be addressed later. While accurate macro measurement is difficult, it is not impossible, and the difficulty need not preclude the implementation of source reduction strategies.

Focusing first on quantity reduction (toxicity reduction measurement issues are discussed later in this chapter), communities can gather basic information about the composition and source of their waste stream, identify trends in population and business growth that affect waste generation, and make choices about how to measure reductions.

Measurement Units: Weight or Volume?

One of the first questions communities establishing a source reduction measurement system need to answer is whether to measure source reduction by weight, volume, or both. As discussed more fully in Chapter 2, weight is most commonly used because the waste stream is usually measured in tons. However, it is volume that is more relevant to landfill capacity and to managing collections by truck. Furthermore, the volume-to-weight ratio of materials varies greatly, as illustrated in Tables 2-6 and 2-7.

Ideally, measuring by both weight and volume would allow communities to assess the tradeoffs when materials are light in weight but bulky in volume, or vice versa. However, this may be costly and difficult for communities that currently measure by weight. At a minimum, when evaluating strategies to reduce waste weight, communities can select those that do not have an adverse effect on waste volume.

Information Needs

Good data collection is vital for measuring source reduction, since communities need to know which sources are generating which types of waste materials, and how much they are generating. Thus, at a minimum, communities need to collect data on:

- Amount of residential waste
- Amount of commercial waste
- Residential population
- Total employment
- Projections of population change
- An index of economic activity

Additionally, information about waste composition, while not essential for overall measurement, allows communities to set appropriate goals (as discussed in the previous section) and identify which materials are being reduced and which are not.

Communities may encounter a variety of problems in obtaining this information or using it to measure source reduction; four key ones are:

1. Distinguishing changes in the overall rate of waste generation from changes in numbers of people or other units generating the waste.

2. Factoring out other external variables, such as the business cycle.

3. Discerning small annual changes when measured by imprecise waste generation data.

4. Separating data for residential and commercial waste (for simplicity, it is assumed that institutional waste is included in commercial waste).

First, changes in total waste generated may result from changes in generating units, changes in generating rates, or both. For residential waste, the generating unit is population. For commercial waste, generating units can be established for each generating sector: number of workers for business waste, number of beds for hospital waste, number of students for school waste, and so on.

Source reduction results must be assessed by identifying a decrease in the generation rate. Source reduction is not necessarily seen just by a decrease in the total waste stream since this may be caused by a decrease in the number of generating units. For example, if a decrease in residential waste generation is accompanied by a decrease in population but no decrease in the per capita generation rate, source reduction has not been accomplished. Similarly, hospital source reduction means less waste generated per bed, not less waste due to a decrease in the number of beds.

Thus, to measure source reduction, a community needs to know whether there has been a change in the magnitude of generating units as well as total waste generation compared to a baseline. Such data are often not readily available and may be costly to collect and analyze.

Second, there are other external variables (besides changes in the number of generating units) that affect waste generation and that cannot be considered source reduction. One example is the level of economic activity. Waste tends to decrease in a recession, totally apart from any source reduction efforts. Seasonal variations can also have an impact, particularly with respect to the amount of yard waste. Thorough measurement would factor out the effects of such variables.

Third, annual reductions in waste generation, if they occur, are likely to be small and must be measured by imprecise waste generation data. Two percent might be a substantial source reduction achievement for one year, but a variation of 2 percent is well within the margin of error in measuring waste generation. To compensate for this, it may be necessary to measure source reduction over longer time periods. In the case of New York State, for example, its 8-10 percent source reduction goal over a 10-year period amounts to less than 1 percent reduction per year. While 1 percent source reduction in a given year may not be measurable, an 8 to 10 percent reduction over 10 years would be much more easily discernable.

And fourth, while separate data for residential and commercial waste are needed to accurately measure source reduction because the generating units are different for the two sectors, many communities do not currently collect information in this way. They may know how much residential waste they generate because collecting and/or disposing of it is a municipal responsibility, but they frequently do not know the amount of commercial waste because this may be collected by private carters. In other localities (New Jersey, for example), the same trucks collect residential and commercial waste and there are no separate data for each sector. Further, much commercial waste is exported from the community in which it is generated. A community cannot know how much waste it generates if it does not know the amount it exports.

Lack of data on commercial waste generation is a major impediment to planning for and measuring source reduction since this sector is estimated to generate 40 percent of municipal solid waste. If the population in a community remains constant but commercial activity increases, there is likely to be an increase in total waste generation even if residential per capita rates do not increase. Conversely, if commercial activity decreases, total waste generation figures would decrease even if residents were not actually reducing the amount of waste they generate on a per capita basis. Thus, if a community knows not just its residential population but also the number of workers, and if it measures residential and commercial waste separately, it can measure source reduction separately — and more accurately — for each sector.

Making Comparisons: Measuring from Baseline Year or Projected Increases?

A key decision for communities is whether to measure reductions from a baseline year or from a projected increase in waste generation. That is, if a goal of 10 percent source reduction from 1990 to 2000 is set, it is important to specify if this means a 10 percent reduction from the 1990 baseline, or a 10 percent reduction from the amount of waste projected for 2000. This decision can have major implications, as Table 3-3 (later in this chapter) shows. Thus, although both methods are valid, it is essential in setting goals and evaluating results to clearly state which method is being used.

Measuring from a projected increase presents a useful assessment of the effectiveness of a source reduction program because it shows the amount of waste that would have been generated without such a program. The drawback of this method is that it is dependent on the accuracy of the projections. If a projection overestimates the amount of waste in 2000, source reduction achievements may appear to be greater than they are. An underestimate in the projections, on the other hand, could make it seem that little or no source reduction had actually occurred. Communities need to take these factors into account when deciding how to measure source reduction.

The American Telephone and Telegraph Company (AT&T) provides an example of how this decision can affect measurements and goals. The company set a goal of a 15 percent reduction in paper waste from the 1990 baseline by 1994. Projected waste was not considered when this goal was set. AT&T now notes that paper waste was projected to increase by 13 percent a year, so achieving its goal will represent a 67 percent decrease from the projected paper waste for 1994. This program is described in Chapter 9.

Measurement Experience

Source reduction has been successfully measured on a "micro" level — on a company or institutional basis, for a small group of households, or for a particular material. Part II of this report includes such measurements for many of the individual strategies described.

For example, for the AT&T office paper waste reduction effort mentioned above, the company estimates potential savings of 77 million sheets of paper per year; this can be measured by paper purchase orders. Similarly, Mount Sinai Hospital in New York City measured source reduction of 25 percent from its switch to reusables in its cafeterias (Chapter 7) by measuring the decrease in number of cans of waste put out for collection. The Minnesota Office of Waste Management has done micro source reduction measurements (in both weight and volume) for the replacement of disposables with such reusables as air filters, drinking cups, and cloth towel rolls, as discussed in Chapter 6.

For the residential sector, a few attempts have been made to measure source reduction, again at the micro level. For example, a study in West Berlin in 1983-1984 measured reduction of 21 percent for 50 volunteer households following an education program.[1] A 1991 Canadian study, in Maxville-Kenyon, Ontario, documented source reduction by 25 families following an educational workshop.

"Dry discards were reduced 34 percent through shopping and behavior changes. Wet discards were reduced 66 percent through backyard composting."[2] Both the Berlin and Ontario studies used volunteers, who may not be representative of the population as a whole, so these results cannot be generalized. However, studies of this type can be extremely useful in estimating the potential for source reduction and evaluating the impact of education programs.

It is the "macro" measurement of source reduction — measuring for a community, state, or the nation as a whole — that has presented major problems. A few attempts have been made, but they have not effectively measured source reduction. For example, Seattle has measured a decrease in the number of cans of trash put out per residence per week from 3.5 to 1.4 since implementing a waste reduction program. Residents are now charged for trash on a per can basis, but recyclables are put out in separate cans, collected without charge, and not counted as trash. Thus, a large part of this reduction is due to separating recyclables and compacting the trash; it is not known how much is due to source reduction. In another example, the town of Perkasie, Pennsylvania, reported source reduction of 26 percent, but there is evidence that some of the reductions were achieved by exporting waste and by burning in home fireplaces[3] — hardly source reduction initiatives.

Solving Measurement Problems

Several approaches can be taken to improve data collection and solve the measurement problems discussed in the preceding sections. The particular measurement techniques used can be decided by each community and will depend on a variety of factors, such as whether source reduction goals are comprehensive or are targeted on specific generating sectors or materials. Another important decision involves determining how much money to invest in data collection and measurement. While measurement is important, there is a trade-off in that money spent for this purpose is not available for implementing programs. The following are some strategies that would aid in measuring source reduction.

- Placing scales at disposal, transfer, and processing facilities could help the data collection process since many communities attribute inadequate data to the lack of scales.

- Imposing variable waste disposal fees can help provide communities with a more accurate measure of residential waste. Some communities are now experimenting with scales on collection trucks that will weigh each household's waste and keep track of this through a bar code label system. Such a system is discussed further in Chapter 11.

- Mandated reporting by private carters on the waste they handle is necessary for establishing a baseline waste generation figure, particularly for the commercial sector. At present, most commercial waste is collected by private carters and communities do not know the total volume of this waste or the amount exported.

- Sampling techniques could be used so that all waste in a community would not have to be measured. Surveys that measure source reduction through backyard composting in Seattle are discussed in Chapter 8.

- Particular materials or segments of the waste stream could be targeted for measurement. For example, shifts from disposable to reusable diapers could be measured by sales of cloth diapers and numbers of diaper service customers. Waste exchanges (discussed in Chapter 8) can document the amount of materials they divert from the waste stream.

❖ Implications of Goal-Setting and Measurement Decisions: An Example

INFORM has been monitoring solid waste management in several New York counties, with a focus on source reduction planning. One of these counties is Dutchess, 100 miles north of New York City. Data from Dutchess County illustrate some of the implications of goal-setting and source reduction measurements.

Dutchess County, which reported generating 250,000 tons of municipal solid waste in 1990, or 5.1 pounds per capita per day, has a 20-year solid waste management plan which states that the county will adopt the New York State goal of 8-10 percent source reduction as its minimum goal. This raises several questions. What are Dutchess County's baseline and target years? How will this source reduction be measured?

The New York State goal is for the 10-year period from 1987 to 1997. Dutchess aims its reduction goal for a 20-year period (from 1990 to 2010), so its goal is actually lower than that of the state. The county has chosen to set a per capita waste reduction goal based on projected increases in waste generation. That is, its plan states that per capita waste generation is expected to increase 8-10 percent by 2010, so preventing this from happening will accomplish 8-10 percent source reduction. Its goal is therefore "no net increase" in per capita waste generation over 20 years.

The choice of which comparisons to make is far from a semantic one. **Table 3-3** illustrates the waste and cost implications for Dutchess County of different levels of source reduction.

Dutchess County projects that per capita waste generation would increase by

Table 3-3 *Waste Generation and Waste Management Cost Estimates for 2010* for Dutchess County, New York*

	10% per capita increase from 1990 baseline* (no source reduction)	10% per capita reduction from projected increase	10% per capita reduction from 1990 baseline*
Total solid waste in 2010	313,000 tons	280,000 tons	256,000 tons
Increase from 1990 baseline	63,000 tons	35,000 tons	6000 tons
Disposal cost increase from 1990 at $150/ton	$9.45 million	$5.25 million	$900,000
Disposal cost increase from 1990 at $300/ton	$18.9 million	$10.5 million	$1.8 million

* 1990 baseline is 250,000 tons, or 5.1 pounds per person per day for 268,600 people; 2010 projection includes a population increase of 37,710 people, or 14 percent. (Source: *Dutchess County Draft Generic Environmental Impact Statement and Solid Waste Management Plan,* March 1991.)

8-10 percent by 2010 if no source reduction program were implemented. This scenario is shown in the first column in Table 3-3, using the 10 percent figure. Waste generation would increase to 5.6 pounds per person per day. With an expected population increase of 37,710 people, or 14 percent,[4] this translates into 313,000 tons in 2010, or an increase of 63,000 tons. At the present average disposal cost of $150 per ton, this would make the 2010 waste disposal budget $9.45 million higher than in 1990.

The Dutchess plan calls for a 10 percent reduction in per capita generation from the projected 10 percent increase, or holding per capita generation constant. This scenario, illustrated in the second column in Table 3-3, would increase total waste by 35,000 tons to 285,000 tons by 2010 (given the projected population increase), and add $5.25 million to the budget, as compared to 1990.

The third column in Table 3-3 shows what would happen if Dutchess County had a real per capita reduction of 10 percent from the 1990 baseline, meaning that, on average, each person would be generating 10 percent less waste in 2010 than in 1990. This would result in per capita waste generation of 4.6 pounds per day in 2010. With the expected population increase, this would increase total waste by only 6000 tons to 256,000 tons and, at the current disposal cost of $150 per ton, add $900,000 to the budget.

All of these estimates assume that current disposal costs of $150 per ton will prevail in 2010. However, these costs are likely to increase substantially over the next 20 years as existing landfills close, export opportunities diminish, and the cost of new facilities increases. The last line in Table 3-3 shows disposal cost increases for each of the three options if disposal costs doubled to $300 per ton. At this rate, the disposal budget would be $18.9 million higher with a 10 percent per capita increase in waste generation but no source reduction, $10.5 million higher with no change in the per capita generation rate, and $1.8 million higher with a real 10 percent per capita decrease in waste generation. At these rates, a real per capita reduction of 10 percent could save the county $17.1 million in disposal costs in 2010 compared to the projected increase without any source reduction.

It should also be noted that Dutchess County does not have an accurate measurement of the waste it generates. As in many areas, commercial waste is handled by private carters and a substantial portion of it is shipped out of the county. To effectively implement and measure source reduction, Dutchess and other communities need to measure all of their waste, including the components that are privately collected and/or exported. Further, use of per capita rates is imprecise since population is the generating unit for residential waste, but not for the commercial sector. In the case of Dutchess County, the per capita waste generation rates were obtained by dividing the total waste generated by the number of people.

❖ Goals and Measurements for Reducing Toxics

The previous discussion of measurement refers to quantity source reduction, not to source reduction of the toxic constituents in the waste stream This presents different and even more complex problems.

It is not possible to set an aggregate percentage reduction goal for toxics in the waste stream, as can be done for waste amounts, for a variety of reasons including: a lack of accurate data on the amounts of toxic constituents, the wide variation in degrees of toxicity, and the difficulty in determining health and environmental impacts. Further, it is difficult to set source reduction goals for individual toxic chemicals in the waste stream, again because of a lack of reliable data.

Thus, the most practical way to set toxicity source reduction goals is to set goals for reducing the purchase and disposal of specific products that are known to contain particular toxic substances. For example, a target might be the number of containers of pesticides, paints, or used oil sent for disposal. Source reduction would be measured as a reduction from the baseline amount on a per capita basis. Chapter 13 explores these issues in more depth.

Notes

1. Marjorie J. Clarke, "The Paradox and the Promise of Source Reduction," *Solid Waste and Power,* February 1990, p. 44.

2. David Allaway, "Does Source Reduction Work?" *Resource Recycling,* July 1992, pp. 59-61.

3. Institute for Local Self-Reliance (Brenda Platt, Christine Doherty, Anne Claire Broughton, and David Morris), *Beyond 40 Percent: Record Setting Recycling and Composting Programs,* Washington, DC, 1990, p. 111.

4. *Dutchess County Draft Generic Environmental Impact Statement and Solid Waste Management Plan,* March 1991.

▓ Chapter 4 Administration and Budget

While source reduction has been hailed as the top solid waste strategy in the US, solid waste departments have traditionally been staffed by officials knowledgeable primarily about waste disposal and, more recently, about recycling. Their responsibilities have been the collection, transport, and disposal of waste, and the processing and marketing of recyclable materials. Their key concerns have been running out of disposal capacity, finding sites for new facilities, and controlling costs.

Implementing source reduction programs involves vastly different staff skills and concerns. It requires staff with a broader, long-term view of the use of materials in society and an understanding of how behavior of consumers, businesses, and government can be changed to optimize the use of materials and minimize the waste generated. Staff members need diverse skills so they can work on planning, program development, technical assistance, education, outreach, legislation, data collection, program evaluation, waste audits, and enforcement. They need information about what is in the waste stream as well as on how much there is, and their concerns encompass broad issues (such as impacts on economic development) that go well beyond questions of how to manage garbage.

❖ Administration

Efforts to provide independence and authority for source reduction are essential if it is to become a viable policy option. For the most effective administrative structure, source reduction would be separate from and independent of waste management functions. The head of the source reduction effort would have authority at least equal to that of the individual in charge of recycling and disposal, and would have a commitment to minimizing the amount of materials actually entering the waste stream.

Independence

Source reduction is much broader in scope than recycling or disposal and is, in fact, resource management rather than waste or materials management. That is, it involves decisions about what products and packages are made, how they are made, and how they are used. An effective source reduction program deals with producers, distributors, and consumers. It can thus be argued that source reduction does not belong in sanitation or solid waste departments at all, and should not be a function of waste managers. Theoretically, it might make more sense to place source reduction activities in a department of economic development. On a more practical level, however, the motivation to promote source reduction is generally the need to reduce waste, so it is likely to remain in the purview of solid waste departments.

If source reduction functions are placed in a solid waste department, they need some independence from the recycling functions because the immediate, every-day demands of recycling can tend to overwhelm the longer term, more complex source reduction activities. Larger budgets and more personnel are required for recycling because it includes collection, processing, and marketing; the scale and urgency of these management tasks may result in eclipsing the attention given to source reduction.

Furthermore, there is some potential conflict of interest between source reduction and recycling. Some communities (the state of New Jersey, for example) give bonuses to recycling departments based on the amount of materials recycled. Source reduction, by cutting the amount of waste generated, can reduce the amount of materials recycled, as well as the amount sent for disposal, thereby reducing these bonuses and the revenues of the recycling department. Recycling income can also be cut by bottle bills and refill programs that reduce the revenues from recycling aluminum cans on which recycling departments often depend.

This conflict is comparable to the conflict between waste disposal and recycling programs, which in effect compete for the materials in the waste stream. Waste disposal staff often count on having given amounts of materials to collect and burn, while recycling staff want some of the same materials for their programs, especially when revenue is derived from sale of collected recyclable materials. A growing recycling program may deprive an incinerator of the tons of waste and related tipping fees it needs on a daily basis. Source reduction can deprive both disposers and recyclers of these materials; hence the need for providing independence for the functions of source reduction, recycling, and disposal.

For all these reasons, recycling departments cannot always be expected to promote source reduction efforts, and it is vital to provide source reduction personnel with sufficient independence that they can focus their efforts exclusively on source reduction.

To some extent, however, source reduction and recycling have a symbiotic relationship. Staff involved in recycling have an awareness of the size and composition of the waste stream. Such awareness can provide a good foundation for promoting source reduction. It is also possible to piggyback source reduction on an existing recycling infrastructure. For example, drop-off centers for the reuse of clothing, books, and equipment can be placed contiguous to recycling drop-off centers. The same company waste audits that lay the foundation for recycling can also be used to identify source reduction opportunities. Cooperative efforts on source reduction and recycling can be promoted by coordinating the two functions while maintaining the desired independence for each.

Authority

Source reduction needs both an advocate at the highest levels of solid waste planning and management, and a director to oversee programs on a daily basis. The director needs status at least commensurate with, if not superior to, that of the directors of recycling and disposal, and a budget to adequately implement a program. Access to top solid waste officials with the broadest overall perspective is essential if keeping materials out of the waste stream is to be the top priority. Such access also ensures that source reduction program proposals do not have to be screened by recycling or disposal officials busy with their own operations.

Current Source Reduction Administration

Despite the virtually universal endorsement of source reduction as the top priority, in no instance did INFORM find an administrative structure that reflects this, in terms of either independence or authority. In fact, INFORM has identified only three localities and two states that have government employees who spend most of their time on source reduction: New York City, Seattle, and Olmstead County (Minnesota), at the local level, and Minnesota and New Jersey at the state level. Often described as waste prevention specialists, most of these individuals are in departments that deal with recycling, spend some of their time on recycling activities, and have very limited resources.

In 1991, the New York City Department of Sanitation (DOS) employed 65 people in recycling, seven of them in planning. At the same time, the source reduction staff was one full-time person, six part-time workers, and three interns and volunteers, all within the Recycling Office. While far from adequate for a city of 8 million people, this put New York out front as a leader since most cities have no distinct source reduction staff.

In July 1992, DOS improved its administrative structure for source reduction by creating a Bureau of Waste Prevention, Reuse and Recycling and elevating its director to the level of Assistant Commissioner. It also established a Unit of Waste Prevention and Related Projects within this new bureau; this unit is to be staffed with four full-time employees. This change represents both a recognition of the importance of source reduction by a city facing a serious solid waste crisis, and a considerable improvement in administrative structure.

Seattle, often cited as the national leader in recycling, is also in the forefront on source reduction, but even in Seattle there is only one source reduction staff member out of twelve in the planning division. The tasks of the others focus on recycling. The source reduction coordinator has spent a large proportion of his time on backyard composting.

Olmsted County, Minnesota (which includes the city of Rochester) has a source reduction coordinator who is responsible for: developing and implementing governmental, commercial, and institutional source reduction programs; providing technical assistance to individuals, businesses, and government agencies; conducting public presentations and media outreach; and researching legislative issues regarding source reduction and packaging. Because of staff constraints within the recycling office (where the source reduction effort is located), the coordinator first worked on business recycling. Despite the administrative chart, there is considerable mixing of source reduction and recycling functions.

The Minnesota Office of Waste Management has three people working on source reduction. One full-time staff member works on technical assistance and business evaluation/education. Another full-time person works on source reduction grants. The third staff member works part time on packaging and shipping research. In addition, Minnesota's Waste Education Program, which focuses on schools and the residential sector, does some source reduction work.

The New Jersey Department of Environmental Protection and Energy has increased the importance of source reduction in the department by creating a Bureau of Source Reduction and Market Development. Within this bureau, one staff member works full time on source reduction, mainly doing business outreach and education. There are approximately 10 other people in the bureau; they work mostly on recycling but do have some source reduction responsibilities. The staff member in charge of yard waste, for instance, works on education for backyard composting and leave-on-lawn programs, as well as on municipal composting.

❖ Budget

Source reduction does not require the costly collection and processing operations involved in waste management options, but it is not free and it cannot be accomplished without an adequate budget. The costs of source reduction are in the form of an up-front investment in data collection, waste audits, legislative development, education, technical assistance, and planning.

The payoff for investment in source reduction is not immediate, but it can be very large, as noted in Chapter 2. In the northeast, for example, the cost of collection and disposal can exceed $200 per ton, and New York City estimates recycling costs at about $300 per ton. In its 20-year solid waste management plan, New York City estimates the cost of some source reduction activities at $20 per ton and avoided costs from source reduction in the $140-$150 per ton range.

Therefore, the plan states, the cost of prevention programs could increase by as much as $120 to $130 per ton before costs would exceed benefits.[1]

New York City's plan further estimates that in the year 2000, if the city reduces 670,000 tons of waste, total avoided waste management costs due to source reduction would be approximately $90 million: $30 million in avoided collection costs and $60 million in avoided processing costs.[2] All of the New York City estimates are based on very modest source reduction of a 7 percent reduction of the projected increase. The New York City plan states that "prevention programs become increasingly cost-effective as prevented percentages increase. The reason for this is that larger prevented tonnages allow relatively greater reductions in truck shifts and facility capacity; conversely, when reductions are smaller, fewer savings are captured through reduced collection and facility costs."[3]

A barrier to funding source reduction is that results may not happen immediately, so there may be no return on the investment in the budget year in which the expense is incurred. In order to assure continual and adequate funding, source reduction could be funded from a designated income stream such as a portion of the funds raised from quantity-based user fees (QBUFs) or other waste collection fees, environmental taxes or fees, or possibly unreturned beverage container deposits. Nineteen states have now placed taxes on items such as yard waste, tires, and lead-acid batteries (see Chapter 11). While the primary purpose of this type of tax is to encourage consumers or users to change what they buy or how they dispose of materials, such taxes could also be used to raise money for other solid waste activities.

Source reduction could also be funded at a specific percentage of a recycling budget. For instance, if 5 percent were chosen, a recycling budget of $10 million would mean a source reduction budget of $500,000.

Current Source Reduction Budgets

A look at source reduction budgets in the three localities and two states identified by INFORM as having staff members spending most of their time on source reduction efforts shows that only small amounts of money have been allocated for source reduction. Further, the amount of money spent for source reduction in states around the country is infinitesimal compared to the amount being spent on recycling.

New York City reported expenditures of $473,000 on source reduction in fiscal year 1991, as compared to $40 million on recycling. **Table 4-1** shows the city's waste reduction and reuse expenses for that year; however, it should be noted that household hazardous waste collection, which the city included in its figures, is not source reduction, but waste management (see Chapter 2).

Seattle funded its source reduction program with $800,000 in 1991, exclusive of the coordinator's salary. Most of this, $564,000, was allocated to backyard composting. Other allocations included $42,000 for a directory of repair stores,

Table 4-1 New York City Waste Reduction and Reuse Expenses, Fiscal Year 1991

Project	Expense
Home composting	$33,300
Waste reduction handbooks	$119,300
Waste reduction posters	$65,000
Canvas/string tote bags	$10,900
Bring your own bag – art	$5,500
Materials for the Arts*	$117,000
Household hazardous waste day	$32,500
Handbook promotion	free
Commercial waste reduction	$20,600
Subtotal	$404,100
Salaries	$69,000
Total	$473,100

* A reuse program, described in Chapter 8.

Source: New York City Department of Sanitation, Bureau of Waste Prevention, Reuse, and Recycling.

$50,000 for a Request for Proposals for a hazardous waste reduction kit, $80,000 for the Environmental Allowance program which gives small grants for source reduction projects, and $20,000 for a waste reduction survey. In 1992, Seattle's source reduction program budget was increased to $871,000. It included new allocations for a retail program and a reuse directory, as well as continued funding for backyard composting, hazardous waste reduction, and grants.

Olmsted County in Minnesota allocated $300,000 to what it labeled "source reduction" in 1989. But almost half was designated for household hazardous waste collection. The remainder was for education, planning, and staff.

The Minnesota Office of Waste Management has a source reduction budget of approximately $150,000 per year. This covers the salaries of the source reduction staff and their miscellaneous office and travel expenses. In 1992, the office was allocated an additional $60,000 for an environmental shopping campaign. New Jersey does not have a separate source reduction budget.

The small amounts of money being spent for real source reduction stand in marked contrast to the growth of funding for recycling. According to a survey by *Resource Recycling* magazine, the average state recycling coordinator administered a budget of $7.2 million in 1991, and 19 percent of the state recycling programs had budgets exceeding $20 million.[4] Key functions listed by recycling coordinators included providing technical assistance, managing grant/loan programs, developing legislation, coordinating local efforts, and education. All of these functions could also be funded to promote source reduction.

The few budget allocations for source reduction described above represent fledgling efforts. Major source reduction results will require much larger budgets and a willingness to think of the long term and have the patience to sometimes wait for a delayed return on investment. The question for communities is not whether they can afford to make the investment but, rather, can they afford not to, given the large potential savings that can accrue from source reduction.

Notes

1. New York City Department of Sanitation, *A Comprehensive Solid Waste Management Plan for New York City and Final Generic Environmental Impact Statement,* August 1992, p. 17.2-2.

2. *Ibid.,* p. 17.2-2.

3. *Ibid.,* p. 17.2-3

4. Jerry Powell, "State Recycling Coordinators: 1991 Salary Survey," *Resource Recycling,* June 1991, pp. 47-48.

▓ Chapter 5 The Abundance of Source Reduction Opportunities

The potential for source reduction is enormous. Virtually every individual and every organization can play a role and become part of the solution to the nation's solid waste problem.

INFORM has identified dozens of examples of source reduction initiatives that are already successfully reducing waste at the source. They come from every sector of US society: state and local government; businesses of all sizes; public institutions such as schools, hospitals, and parks; citizens' groups; individual consumers; and nonprofit organizations.

INFORM chose the specific initiatives highlighted here because they have the potential to be adopted in other communities, or to be adapted to other situations. For example, businesses and government agencies could adapt programs that reduce school cafeteria waste to food service facilities in their own organizations. Effective state legislation can serve as a model for city legislation. Government agencies can adopt business initiatives and vice versa. Government waste educators can glean additional ideas from consumer education programs developed by a citizens' group, as well as from other government-sponsored programs. The possibilities are practically limitless.

Part II organizes the strategies identified by INFORM into eight categories, for ease of reference.

1. Government source reduction programs (procurement and operations) (Chapter 6)

2. Institutional source reduction programs (Chapter 7)

3. Government assistance programs (technical assistance, backyard composting and leave-on-lawn assistance, grants, pilot programs, clearinghouses, awards and contests, and reuse programs) (Chapter 8)

4. Business source reduction programs (Chapter 9)

5. Education (in households and in schools) (Chapter 10)

6. Economic incentives and disincentives (variable waste disposal fees, taxes, deposit/refund systems, tax credits, and financial bonuses) (Chapter 11)

7. Regulatory measures (required source reduction plans, labeling, bans, and packaging) (Chapter 12)

8. Programs to reduce the amount of toxic chemicals entering the municipal solid waste stream (Chapter 13)

The text includes explanations of each of the strategies, as well as examples of communities or organizations that have tested them. Where possible, contact names are listed with the discussion of specific source reduction initiatives, to give planners access to additional resources for designing and implementing their own source reduction programs.

Chapter 6 Government Source Reduction Programs

Government — federal, state, or local — employs one out of every six workers in the United States, a total of 19 million people. Successful efforts to reduce the waste generated by this work force could not only have a great impact on the municipal solid waste stream, but could also provide a model for businesses, institutions, and consumers.

Exact figures on how much waste government workers produce are not available. However, given that the nonresidential sector generates about 72 million tons of waste each year,[1] it can be estimated that government generates approximately 12 million tons of waste annually. As part of this total, government agencies generate almost 1.7 million tons of office paper waste each year, over 20 percent of all paper waste from offices throughout the country.[2]

Strategies for implementing source reduction in government agencies (as well as other businesses) fall into two main categories: (1) changing procurement policies and (2) modifying operations.

❖ Procurement

Government purchases of goods and services at the federal, state, and local levels account for approximately 20 percent of the gross national product,[3] or about $1 trillion a year. Hence, government as a whole has great procurement power. Changes in government procurement policies to favor source reduction could have impact both within and beyond government.

While the federal government and many states have adopted procurement policies favoring recycled goods, these agencies have rarely used procurement policy to achieve source reduction objectives. Yet, procurement guidelines could require the purchase of reusable, refillable, repairable, more durable, and less toxic items. They could also require minimal and reusable packaging.

Clearly, if governments purchase less wasteful and less toxic products, they will create less waste. Further, procurement policies favoring source reduction could save money for governments in many ways, including purchasing, mailing, and disposal costs. Considering disposal alone, INFORM has estimated the average national disposal cost to be $100 per ton. Using the estimate of 12 million tons of waste generated annually by government, the annual cost of government solid waste disposal would be about $1.2 billion a year.

The impact of government procurement policies favoring source reduction would also extend beyond government itself. First, government procurement power could encourage manufacturers, shippers, and retailers to develop products and packages that generate less waste and contain less toxic constituents — products and packages that would then be available to all purchasers. Second, a government source reduction procurement program could be a model for the private sector. Two examples of private sector procurement initiatives are detailed in Chapter 9.

Twenty of the 27 states surveyed by INFORM have some policies pertaining to the purchase of materials with recycled content, but only five have procurement programs aimed at reducing waste generated: Connecticut, Illinois, Minnesota, Rhode Island, and Wisconsin. (Indiana, not one of the 27 states, also has a source reduction procurement program.)

Connecticut

■
Connecticut Department of Administrative Services
Contact
Peter Connoly,
Director of
Purchasing
Address
460 Silver St.,
Middletown, CT
06457
Phone
203-638-3267
■

Connecticut is the only state that has legislation requiring state agencies to take steps to eliminate products that are not reusable: Public Act 89-385, passed in 1988. While the act does not mandate specific actions, it does list opportunities for purchasing changes that would reduce the government's reliance on disposable products. The products recommended for purchase are:

- refillable ball point pens
- multistrike ribbons for typewriters and printers
- refillable laser toner cartridges
- mechanical pencils
- reusable razors
- recycled cloth rags instead of disposable wipers for garages
- cloth diapers for appropriate agencies
- reusable washable aprons, cafeteria hats, and tablecloths
- washable tableware (to be implemented where dishwashing equipment already exists)
- food such as dressings, condiments, and juices in large reusable packaging containers
- re-refined oil for use in fleet vehicles

The act also encourages purchasers to:

- buy certain items (such as detergents and cleaners) in bulk, in large institutional-type containers
- eliminate contracts for disposable food service paper products
- issue contracts for retreaded tires for truck and passenger vehicles
- educate employees to reuse paper clips, rubber bands, and brass fasteners
- require agencies to use only reusable envelopes for interdepartmental agency mailings
- require vendors that deliver on pallets or in 55-gallon drums to be responsible for the exchange of these items on subsequent deliveries

The Connecticut Department of Administrative Services reports that the switch to reusables has been a slower process than planned for two main reasons.[4] First, reusables often cost more initially than disposables. Some reusable products that require high start-up costs are reusable tableware, cloth diapers, and the equipment needed for cleaning. Connecticut has a large state deficit, so state purchasers are trying to avoid buying items that cost more, even if savings will result in the long run. Individual agencies, which have some purchasing authority for commonly used items, also try to save money by buying the less expensive disposables. Second, no enforcement mechanism is built into the legislation to prevent agencies from purchasing disposables.

The Connecticut Department of Administrative Services has not yet calculated the cost savings or waste reductions achieved by its source reduction procurement policies, but state purchasers have made some general observations about the switch to some of the reusable items. For example, they report that refillable pens and multistrike ribbons are more expensive initially than their single-use counterparts, but result in future cost savings; in addition, refillable laser toner cartridges are saving money and seem to be functioning as well as the previous disposables. However, the agencies report, the purchase of multistrike ribbons by individual agencies has decreased since the beginning of the program because the ribbons produce poorer quality type. Since the preference seems to be for disposable film typewriter cartridges, government might be able to use its purchasing power as a major consumer to encourage the ribbon-makers to refill the plastic cartridges with new ribbons.

Illinois

Illinois has adopted procurement policies that mandate government agencies to use double-sided copying, routing slips to share publications, and soy-based inks for printing.

In addition, the Office of the Illinois Secretary of State uses two-way envelopes to send license plate renewal applications. Two-way envelopes have an extra flap with the return address that can be refolded to make a new envelope. With these envelopes, it is unnecessary to include return envelopes and, therefore, 25 to 30 percent less paper (by surface area) is used. Since the office sent out almost 8 million two-way envelopes in 1990, almost 8 million return envelopes were never purchased or disposed of, and the office saved more than $57,000 in purchase costs. Two-way envelopes also require less storage space.[5]

Minnesota

A 1980 state statute required the Minnesota Department of Administration to develop and implement policies to recycle and reduce waste by changing what it purchases. This program now focuses on source reduction.

The department is producing a guidebook and training program for state officials to teach them about source reduction. It has also completed an assessment of energy usage of certain equipment and plans to start an analysis of cost versus waste issues for state purchases.

Within the Department of Administration, the Materials Management Division is preparing a procurement model and routinely reviews its purchasing criteria to encourage the purchase of durable, reusable, and repairable items, as well as to increase its purchases of recyclable and recycled materials. The division is pursuing procurement aimed at source reduction through four strategies: increasing purchases of reusable items, buying durable products, purchasing remanufactured goods, and buying used or surplus equipment.

With regard to increased purchases of reusable items, contracts have been signed for:

- buying refillable instead of disposable pens
- renting and cleaning laboratory garments rather than buying disposables
- refurbishing highway signs instead of disposing of old or vandalized ones

The division is promoting source reduction through increasing purchases of durable goods by requesting extended warranties for certain purchases. For example, the office furniture contract specified a 10-year warranty, rather than the 3- to 5-year warranties normally offered in the industry. The Department of Administration orders approximately $2 million worth of furniture under this contract annually.

As one example of purchasing remanufactured products, the department has a contract with a company to retread tires from radial and off-road truck tires, which amounts to an estimated 582 truck tires per year. Using an average weight of 80 pounds per truck tire, the estimated annual source reduction achieved by this measure is 23 tons of waste.

■
Minnesota Department of Administration, Materials Management Division, Resource Recovery Office
Contact
Lynne Markus, Manager
Address
112 Administration Building,
50 Sherburne Ave.,
St. Paul, MN 55155
Phone
612-296-9084
■

To increase purchases of used and surplus equipment, the fourth program area, the department operates a distribution center in Arden Hills, Minnesota, where state and federal surplus items are displayed and offered for sale. During 1990, 67 state agencies, 210 political subdivisions, and 89 school districts purchased hundreds of used desks, chairs, filing cabinets, bookcases, typewriters, and desktop supplies otherwise destined for landfills. In just two of these purchases, state agencies bought 403 used desks and 1094 used chairs. These purchases alone reduced the waste stream by 77.73 tons.[6]

■
Rhode Island Solid
Waste Management
Corporation
Contact
Erica Guttman,
Environmental
Program Planner
Address
West Exchange
Center,
260 West Exchange
Street,
Providence, RI 02903
Phone
401-831-4440
■

Rhode Island

The Rhode Island Solid Waste Management Corporation (RISWMC), a quasi-state agency, wrote a report in 1991 on how to change procurement practices to reduce solid waste generated in state government offices. Based on the agency's research and cost/benefit analyses, the report made the following recommendations for Rhode Island government agencies. (The report did not attempt to factor in complete lifecycle analyses.)

- Promote increased use of state and federal surplus property programs (buying surplus furniture and other supplies at lower than retail prices) through education and by requiring state agencies to send requests to surplus property officers; the surplus property is stored in warehouses and, if not used, may eventually be disposed of as waste.

- Conduct a pilot program with permanent air filters (that can be cleaned and reused) and cloth towels in restrooms to determine cost effectiveness.

- Refill typewriter ribbon and laser printer cartridges.

- Buy copiers with duplex capabilities (ability to make two-sided copies) when replacing old copiers.

- Use pre-printed post cards to get off unwanted and duplicate mailing lists and use "two-way envelopes" (envelopes that can be refolded and reused) for return mailings .

- Ban the purchase of disposable cups.

The report also stated that the state could promote source reduction further through legislation, education, and a source reduction requirement in the bidding process.[7]

The RISWMC report recommended that government agencies interested in implementing source reduction procurement policies first conduct pilot programs. Results from the pilot programs could then be assessed by the governor's environmental policy staff, and an executive order could be given to government purchasers to carry out certain procurement changes.

Due to current staffing constraints, RISWMC has only been able to conduct a pilot project comparing paper towels to cloth towels. Cloth towels were put in restrooms on two different floors in a state building for one month. The use of cloth towels in these restrooms reduced paper towel waste by more than 85 pounds per week, a total of more than 342 pounds.[8] However, the results also showed staff opposition to the cloth towels. On one of the floors, where the restrooms were also open to the public, more than 88 percent of the workers on the floor thought the use of the cloth towels was unsanitary and did not like using them. On the other floor, 65.7 percent of the staffers opposed the cloth towels even though the restroom was mainly used by state employees. Nevertheless, RISWMC is still committed to promoting reuse strategies in state agencies and will attempt the cloth towel and other pilot projects in other state offices when staffing allows.

Wisconsin

Although Wisconsin has had legislative requirements to determine lifecycle costs of products purchased since 1982, it has only recently begun to fulfill them. Lifecycle costing is helpful in comparing the costs of durable and reusable products with the costs of disposables because it assesses the cost of products over their useful life and thus can be used to estimate costs per year. More durable products generally cost more initially, but the annual cost over their lifetime may be lower than that of disposables.

The state procurement recycling coordinator is developing cost assessment formulas for the products agencies buy. For example, the formula for lawnmowers will include fuel use and durability. The switch to equipment that will generate less waste will be facilitated by the completion of the formulas.

According to a state purchaser, there has been little movement so far towards purchasing reusables in new product areas. An exception is that many municipal hospitals have switched to cloth diapers from disposables. The state plans to examine the cost effectiveness of this program and encourage state hospitals to make the change if it is found to be cost-effective.

The Wisconsin Department of Administration currently has an electronic bulletin board that other purchasers, businesses, and vendors can call up to see a list of recycled products. The list is frequently changed and updated. The lifecycle analyses will be added to the computer directory when they are completed and will provide an opportunity for purchasers (in government and private industry) to evaluate products with respect to source reduction as well as recycling.

Other Source Reduction Procurement Possibilities

Since source reduction procurement is not yet commonplace, many possibilities have not been tested. INFORM has identified some procurement policies that may deserve consideration as good candidates for achieving source reduction.

■
Wisconsin Department of Administration, Bureau of Procurement
Contact
Dan Wehram, Procurement Recycling Coordinator
Address
PO Box 7867
Madison, WI 53707
Phone
608-267-6922
Computer Bulletin Board
608-267-2723 (available with communications software and modem)
■

- Set a price preference for reusables, refillables, durables, and equipment that reduces waste, such as double-sided copy machines. Price preferences, which are commonly used to encourage the purchase of recycled or recyclable products or to promote other social goals, allow purchasers to pay 5 to 10 percent more for specific products. In this case, the price preference would be designed to encourage purchase of products that create less waste but that may cost more than similar products. One drawback to this strategy, as reported by the Minnesota Department of Administration, is that some companies add the 5 to 10 percent on to their prices if there is a preference.[9]

- Require companies that ship goods to government agencies to package them in reusable shipping containers and/or to take back the containers or packaging. For example, furniture can be delivered in reusable shipping blankets.

- Require suppliers to eliminate unnecessary layers of packaging.

- Require government agencies to ship materials in reused and reusable packaging.

- Publicize source reduction procurement guidelines through trade purchasing organizations. The National Association of Purchasing Management, Inc. is a not-for-profit association that provides purchasing information to more than 35,000 members.

- Negotiate for longer and more comprehensive warranties and service contracts when purchasing durable goods.

- Consider leasing equipment instead of buying it to provide manufacturers with an incentive to keep it in good repair.

- Purchase laser printers that can make double-sided copies, computer software that permits faxing from a computer to reduce printouts, fax machines that use plain paper (this reduces the tendency to photocopy and then discard fax paper), narrow-lined note pads, reusable coffee filters, refillable tape dispensers, longer lasting light bulbs, and sturdy desk supplies such as book ends and file holders. Most of these items are available through existing office supply stores.

■
National Association
of Purchasing
Management, Inc.
Address
P.O. Box 22160,
Tempe, AZ
85285-2160
Phone
602-752-6276
■

❖ Government Operations

Changing government operations to promote source reduction could contribute to lessening both the amount and toxicity of waste. It could also save government money and serve as a model for the private sector. For example, government offices with lawns and campuses can compost yard waste on site and leave or mulch grass clippings on lawns. Employees can be educated to reduce paper use and reliance on disposables, and to reuse materials that they might otherwise discard, such as paint.

Government agencies can begin source reduction on their own, or they may be mandated by a mayor or governor to implement source reduction programs. A recent example of this kind of mandate occurred in New Jersey. An executive order signed by Governor James Florio in June 1991 that required government agencies to conduct waste audits and develop plans on how each agency will meet the state's 60 percent recycling goal also included certain specific source reduction requirements:

- Departments with public lands must leave grass clippings on the lawn.

- Each office must provide a copy machine capable of making two-sided copies.

- Departments must replace disposables with reusables, where feasible.

INFORM's research did not find any examples of extensive changes in government operations that accomplish source reduction. However, a pilot program being conducted in Itasca County, Minnesota, started to make such changes; it is discussed in Chapter 8. Leave-on-lawn programs, which can be applied to government property, are also described in Chapter 8. In addition, INFORM has calculated the potential cost savings (in purchase and disposal) associated with reducing office paper use. A corporate program (by AT&T) that has achieved significant savings in this way is described in Chapter 9.

Office Paper Reduction

While INFORM did not identify any major government programs in this area, office paper is an excellent candidate for source reduction. It is an important segment of the waste stream, and organizations have a relatively high degree of control over its use and disposal. Office paper is one of the fastest growing segments of municipal solid waste in the United States. In 1960, 1.5 million tons were generated, or 1.7 percent of municipal solid waste. This rose to 7.3 million tons in 1988 (4.1 percent of municipal solid waste) and is projected at 16 million tons by 2010 (6.4 percent of municipal solid waste).[10]

Government's use of office paper comprises over 20 percent of all office paper used in the country, or 1.5 million tons in 1988.[11] Thus, office paper reductions made in government agencies could greatly reduce the waste stream.

If offices across the country increased the rate of two-sided photocopying from the current rate of 26 percent to 60 percent, and reduced the amount of copies made by one-third (from 446 billion to 297 billion[12]), close to $1 billion a year could be saved in avoided disposal costs and paper purchases, according to INFORM's calculations.[13] This would save 890,000 tons of paper annually and over 15 million trees.[14]

Some paper reductions can be achieved solely by increasing double-sided (duplex) photocopying. Even greater reductions can be made by also reducing the number of copies made and increasing the intensity of use. A document double-spaced and single-sided uses four times as much paper as a document that is single-spaced and double-sided (duplexed). Some strategies for reducing paper and waste include:

- copying on two sides
- using central bulletin boards
- posting or circulating documents
- eliminating fax cover sheets
- editing and carefully proofreading on the computer before printing
- setting up central filing systems
- storing files on computer disks
- refolding and reusing file folders
- reusing envelopes with metal clasps
- loading laser printer paper trays with paper used on one side for drafts
- reducing direct mail by targeting audiences as narrowly as possible
- using small pieces of paper for short memos
- using paper that has already been used on one side as scrap or note paper
- setting narrower margins for drafts
- changing margins to avoid pages with little text
- single-spacing documents

Additional opportunities for paper reduction and other source reduction initiatives that can be taken within offices are:

- encouraging employees to reuse lunch bags
- sharing newspapers and magazines
- organizing swaps of items from home for employees, and providing sections in company newsletters and spaces on company bulletin boards where employees can list information about items available for swapping

Notes

1. Personal communication: Nicholas Artz, Franklin Associates, Ltd., to Caroline Gelb, INFORM, March 1992. Franklin estimates that the commercial sector generates 40 percent of the national municipal solid waste stream by weight.

2. Franklin Associates, Ltd., *National Office Paper Recycling Project: Supply of and Recycling Demand for Office Waste Paper, 1990 to 1995,* July 1991, p. B-19.

3. *Statistical Abstract 1990,* p. 425, and "Government Purchasing Project," testimony by Ralph Nader, Eleanor J. Lewis, and Eric Weltman delivered to the Subcommittee on Oversight of Government Management of the Senate Government Affairs Committee, November 8, 1991, p. 1.

4. Information for this section was provided by Peter Connoly, Director of Purchasing, Connecticut Department of Administrative Services, in personal communications with Caroline Gelb, INFORM, March 4, 1991 and October 17, 1991.

5. Minnesota Office of Waste Management, *Examples of Source (Waste) Reduction by Commercial Businesses,* March 1989, p. 3.

6. Minnesota Department of Administration, Materials Management Division, *Resource Recovery Biannual Report: FY 1989 and 1990,* submitted to the Legislative Commission on Waste Management, January 1991.

7. Rhode Island Solid Waste Management Corporation, *Report on Waste Reduction in Rhode Island State Agencies,* August 1991.

8. *Ibid.*

9. Personal communication, Lynne Markus, Manager, Resource Recovery Office, Minnesota Department of Administration, Materials Management Division, to Caroline Gelb, INFORM, March 16, 1991.

10. US Environmental Protection Agency, *Characterization of Municipal Solid Waste in the United States: 1990 Update,* Washington DC, June 1990.

11. Franklin Associates, Ltd., *National Office Paper Recycling Project: Supply of and Recycling Demand for Office Waste Paper, 1990 to 1995,* July 1991, p. B-19.

12. Banking Information Systems (BisCap), Norwell, MA (a consultant to the office machine industry).

13. INFORM (Robert Graff and Bette Fishbein), *Reducing Office Paper Waste,* New York, November 1991, p. 16.

14. National Wildlife Federation's "Citizens Action Guide" estimates savings of 17 trees for every ton of paper used.

▓ Chapter 7 Institutional Source Reduction Programs

Institutions — organizations such as correctional, health care, educational, and cultural facilities — can also play significant roles in reducing the municipal solid waste stream. While the amount of US solid waste generated by institutions is not known, a study by CalRecovery, Inc. (a consulting firm) estimated that institutions generate approximately 11 percent of the municipal solid waste in New York City.[1] Local governments could advance institutional source reduction by implementing programs in the facilities they operate and by encouraging privately run institutions to replicate these efforts. A variety of source reduction programs, some of which are highlighted here, are currently being implemented in correctional facilities, hospitals, and schools around the country.

❖ Correctional Facilities

Correctional facilities on the local, state, and federal levels present major opportunities for governments to reduce waste, since correctional officials have a high degree of control over how much and what kind of waste is produced. According to the US Census Bureau, there were 856,332 prison and jail inmates in 1987. Based on a waste generation rate of 4 pounds per prisoner per day (the estimate used by New York State prisons),[2] prisoners throughout the country may generate more than 625,000 tons of waste each year.

New York State Prisons

New York State implemented an extensive prison source reduction and recycling program in 1990.[3] Prior to this, some 60,000 state prisoners generated approximately 44,000 tons of waste each year.

To target materials as the best candidates for reduction, the Department of Correction (DOC) conducted a waste survey in the 68 state prisons. DOC found the prison waste stream to consist of approximately 25 percent corrugated

■

New York State Department of Correctional Services
Contact
James I. Marion, Recycling Program Manager
Address
Sullivan Correctional Facility,
PO Box AG,
Fallsburg, NY
12733-0116
Phone
914-434-2080
■

cardboard (half of which was from food delivered to kitchens), 25 percent organic waste (mostly kitchen scraps, not leftovers), and 15 percent tin, with the remaining 35 percent office paper, newspaper, styrofoam, plastic containers, and miscellaneous materials such as wood and scrap metal.

Based on this survey, the DOC implemented five source reduction strategies within the state's prisons: on-site composting of food and cotton mattresses, switching to reusables, eliminating trash bags, purchasing goods in bulk, and reducing paper use. Approximately 30 state prisons have instituted comprehensive programs.

Through these programs, DOC estimates achieving a 65 percent level of waste reduced or recycled at these facilities; DOC did not have separate figures for source reduction. The department reported saving $1.2 million in avoided disposal costs from October 1990 to October 1991, and estimated that disposal costs have been cut in half at the 30 facilities with comprehensive programs. Contributing to the success of the program is the fact that the labor cost for composting, washing reusable tableware and trash cans, and other activities involved in running the programs is low since prisoners are paid only $1.50 a day. The program is broad in scope and progress is well tracked because every prison must report results to a single state-level recycling coordinator each month.

On-site composting began in the fall of 1990. Thirty facilities now compost their wastes, and an additional 25 facilities are expected to be composting when the program is fully implemented (the remaining 13 prisons have land or space restrictions). DOC is developing regional composting sites so the facilities that cannot compost on-site can transport organic materials to a collective compost site.

DOC estimates that one pound of food waste is produced per prisoner per day. The 60,000 prisoners in the system, therefore, generate 30 tons of food waste each day. In October 1991, 385 tons of organic material were composted. With tipping and hauling fees of $115 per ton, the prison system saved more than $44,000 in avoided disposal costs in that one month. The $20,000 cost of the composting equipment was thus offset by avoided disposal costs in less than a month.

The Department of Corrections uses most of the compost within the prison system at its eight farm sites or in prison horticulture vocational training facilities. It donates about 1 percent to community service programs or to prison workers who take it home. The prisons receive yard waste from the state Department of Transportation and local landscapers to use as a bulking agent in the composting process.

DOC also composts the cotton filling of mattresses (it replaces about one-third of all prison mattresses each year). Prisoners strip the mattresses and compost the filling in 15 facilities. Since there are about 1000 prisoners in each of these facilities, or a total of 15,000 prisoners (and, therefore, 5000 mattresses replaced each year), and since each mattress weighs 20 pounds, approximately 100,000 pounds (or 50 tons) of waste are diverted from disposal each year at just these 15 facilities.[4] Since disposal costs are $115 per ton, these prisons are saving $5750 in avoided disposal costs.

The DOC is also beginning to switch to reusables in food service. Approximately 12,000 prisoners are "keeplocked" — isolated for medical, safety, or disciplinary reasons. They receive all meals in disposable styrofoam trays in their cells. Since they receive three meals a day, over 13 million disposable trays are thrown out each year. A complete return to reusables could reduce much of this waste.

A prison study found a polycarbonated plastic tray (trade name Lexan) to be the best option for eliminating the styrofoam trays: it is unbreakable and cannot be used as a weapon. Each Lexan tray costs $22 and is expected to last 4 to 5 years, compared to single-use styrofoam trays that cost $0.02 each.

The Lexan trays are just beginning to be used in the prison system and there are no figures available as to how much waste the entire prison system is avoiding. One prison with 450 prisoners, all keeplocked, has switched entirely to reusable Lexan trays. These prisoners formerly used 492,750 styrofoam trays a year, which cost $9855. The total cost for 450 reusable trays was $9900. Therefore, the cost of the reusables could be recovered in one year. If the trays last 3 more years, the prison could save almost $30,000 in purchasing costs over this period.

When this Lexan tray program is expanded to the entire prison system, about $800,000 a year could be saved in purchasing costs over the 4-year lifespan of each tray, in addition to savings in disposal costs. Prison costs not in this analysis include labor for washing the trays, and costs of cleaning supplies and additional energy used.

A third DOC program eliminates trash can liners. People who work in administrative offices are encouraged to avoid trash bags, since most waste generated is dry. To eliminate trash bags in cans that hold wet waste (such as food), prisoners are assigned to wash trash cans. No figures are yet available regarding the amount of waste reduced and purchasing costs avoided.

The fourth DOC source reduction program involves buying food in bulk. In order to save money and reduce waste, the state prison system is beginning to centralize its food processing facilities. Currently, most food is prepared in individual prisons. Under the new system, food will be prepared in one of four centralized facilities throughout the state and shipped chilled to individual prisons to be heated up and served. The first centralized kitchen opened in April 1992 and processed food for 17 prisons; DOC planned to serve 25 prisons by September 1992. With the centralized system, food can be bought in bulk, reducing tin and cardboard in the waste stream as well as saving money in purchasing costs. The environmental impact and cost of the additional transportation, and the waste associated with shipping to individual facilities, have not been assessed.

The move to bulk purchasing is already showing waste reductions. One facility switched from buying individual serving cereal boxes and half pints of milk to buying both in bulk. This change cut waste from these sources in half. Other items, such as tomato sauce, will also be purchased in bulk refillable containers.

The fifth source reduction program currently underway at DOC facilities is encouraging paper use reduction through a variety of activities. For example, some prisoners bind used computer printout paper to make notepads for classes. About 20 prisons are contributing low-grade office paper and newsprint to prison farms to be used as animal bedding (thereby avoiding the need to purchase and dispose of new bedding material). Information on the number of tons of waste avoided and on avoided disposal costs is not yet available, but DOC reports there will be savings in purchasing costs since animal bedding costs $35 per ton.

❖ Hospitals

The United States Census Bureau listed 1,261,000 hospital beds in the United States in 1987. The Tellus Institute estimates that the waste generation rate is 3.2 tons per hospital bed per year.[5] (These per bed rates incorporate all the solid waste in the hospital, including that generated outside the patient care areas, divided by the number of beds.) Thus, at full occupancy, hospitals alone would generate approximately 4 million tons of waste per year, which is more than 2 percent of the national waste stream. New York City estimates that all health care facilities (including hospitals, clinics, doctors' offices, and veterinarians' offices) comprise 4 percent of the city's total waste stream.[6]

The composition of health care facility waste streams differs depending on the type of facility (such as acute care facilities and offices), but all facilities have regulated medical (infectious) waste (RMW) and nonregulated medical waste (NRMW). RMW is disposed of in red plastic bags. It is generally referred to as "red bag waste" and NRMW is called "black bag waste." In New York City, for example, approximately 75 percent of total health care waste is handled as black bag waste and the remaining 25 percent as red bag waste.[7] But, only one-quarter of what is handled as red bag waste is actually RMW — items such as intravenous bags, tubing, and sharps (needles, syringes, and other medical tools), medical or laboratory apparatus containing body fluids, and pathological and blood-soaked waste.[8] The remaining three-quarters is actually NRMW — items such as newspapers, flowers, and pizza boxes from patient rooms that are put in red bags through error or fear of liability. This is a very costly practice since red bag waste costs at least 15 times as much to dispose of as black bag waste.[9] While the New York City figures may differ from those around the country, they provide an example of how waste collection practices affect disposal costs.

Black bag waste is generally disposed of in the municipal solid waste stream, while red bag waste is subject to more stringent disposal regulations. Most red bag waste is incinerated, and there are current efforts around the country to sterilize and recycle the plastic components. New York City ships most of its red bag waste out of state. Major cost savings could be achieved by keeping NRMW out of the red bags; however, this would not reduce the amount of waste actually generated.

There are, however, many opportunities for accomplishing source reduction of nonregulated medical waste. Below are some examples of source reduction measures implemented throughout the country that could be duplicated by other health care facilities.

New York City Health and Hospitals Corporation

■
New York City
Health and Hospitals
Corporation
Contact
Dexter Dugan,
Senior Management
Consultant
Address
230 West 41st St.,
New York, NY 10036
Phone
212-391-7402
■

The New York City Health and Hospitals Corporation (HHC) is the largest health care institution in the nation. At its eleven acute care hospitals, five long-term care facilities, and five neighborhood family care centers, it has more than 10,600 patient beds and 50,000 employees.[10] HHC recently conducted a citywide medical waste management study and has developed a waste management plan for all medical waste generators within New York City, including site-specific plans for all HHC facilities. As of early 1992, it had implemented or considered a variety of specific strategies.

Strategies completed:

- Switched from disposable corrugated cardboard boxes and red bag liners to a reusable container to hold and ship RMW for disposal.

As a result of this measure, HHC expects to see a 3.4 percent reduction in its total waste stream, reducing over 1250 tons per year. The new contracts require carters to supply and clean the reusable containers.

Strategies in process:

- Replacing paper towel dispensers with hot air dryers.
- Reducing use of disposable linens and disposable food service items in patient rooms.

These measures are expected to reduce HHC's waste by 6.7 percent — nearly 2500 tons annually.

Strategies planned:

- Establish an HHC corporate Product Packaging Evaluation Committee. The goal of the committee will be to change purchasing practices and warehouse product handling procedures to reduce waste. The committee will also be responsible for developing reduced packaging criteria.
- Convert cafeterias to reusable tableware.

Table 7-1 *Estimated Annual Source Reduction Savings for a 1000-Bed Hospital**

	Savings	
Strategy	Tons per year	$ per year
Replace paper towels with air dryers	100	$45,000
Replace disposable food service items with reusables	200	$500,000
Eliminate use of plastic trash can liners in administrative areas	7	$20,000
Replace disposable linens with reusables	150	$200,000
Replace disposable admissions kits (water pitchers, glasses, and bed pans) with reusables in patient rooms	20	$150,000
Switch from disposable to reusable containers for sharp medical instruments	17	$175,000

* Savings include washing, laundering, and other service costs, but do not include capital costs.

Source: New York City Department of Sanitation, A Comprehensive Solid Waste Management Plan for New York City and Final Generic Environmental Impact Statement, *August 1992, pp. 7-14, 7-15.*

The HHC study also estimated the cost and waste savings from selected source reduction strategies for a 1000-bed hospital, as shown in **Table 7-1.**

The HHC study also estimated that assigning costs to departments based on the amount of waste they generate could reduce waste generation up to 20 percent, with savings of $500,000 annually. In addition, the study cited the need for improved control of the discarding of unused products. Improper ordering methods and hoarding often result in perishable products becoming outdated and thrown out before being used. A 10 percent reduction in the number of unused products discarded would save about $100,000 annually for a 1000-bed hospital.[11]

Hospital of Saint Raphael

■
Hospital of Saint Raphael
Contact
Phil Radding
Address
1450 Chapel St., New Haven, CT 06511
Phone
203-789-3139
■

The Hospital of Saint Raphael, in New Haven, Connecticut, developed a comprehensive waste reduction program through a committee that was designed to focus on waste issues.[12] Saint Raphael is a 491-bed hospital that cares for more than 20,000 in-patients a year and has a staff of 3300. The committee designed a "Cash for Trash" campaign which encouraged employees to submit waste reduction ideas in exchange for a cash bonus or prize value of $15. As a result, several source reduction strategies were suggested and implemented in the general areas of reducing paper use and switching to reusables. Complete cost savings and waste reduction information is not yet available.

Reducing paper use:

- Increasing double-sided copying in the central copy area. This has resulted in an approximately 40 percent reduction in paper usage.

- Reducing amount of faxing through an education campaign. The hospital also has a number of plain paper fax machines that reduce paper use by eliminating frequent copying of documents on fax paper.

- Reusing materials in the in-house day care program. The children mainly draw on the back of used office paper and are asked to bring used paper and yarn from home to do arts and crafts projects.

Switching from disposables to reusables:

- Selling plastic reusable mugs with no-spill lids and the Saint Raphael logo to staff for $1. Staff members receive a discount on hot beverages in the cafeteria when they use these mugs.

- Encouraging the use of reusable tableware at special functions, such as meetings and conferences. The hospital has a full supply of reusable tableware for approximately 70 people. Dishwashing equipment is available since patients already receive most meals on reusable tableware.

- Buying most cleaning substances in 55-gallon drums that are refilled by the supplier.

- Switching from disposable bed pads to reusables.

- Encouraging staff to use fewer paper napkins, take disposable flatware only as needed, use less paper, and use stainless steel instruments that can be sterilized, reused, and eventually recycled as part of a "Think Before You Use" campaign.

Other source reduction efforts that have been implemented include:

- Cutting down on the amount of multiple medical publications received by asking all staff to share journals and magazines.

- Encouraging staff to bring in magazines from home to distribute to patients.

The committee did not think that all the suggested source reduction strategies would work. Strategies they had doubts about were:

- Switching to refillable salt and pepper shakers (many would be stolen).

- Returning to reusable tableware in the cafeteria (doctors and nurses might bring food back to offices and the dishes/cutlery would be lost or create unsanitary conditions). One possible solution to this problem, identified by INFORM, could be to have collection locations near staff offices.

■
Minnesota Office of
Waste Management
Contact
Kenneth Brown,
Waste Reduction
Management
Address
1350 Energy Lane,
St. Paul, MN 55108
Phone
612-649-5749
800-652-9747
(toll free in MN)
■

■
Mount Sinai Hospital,
Support Services
Contact
Michael Connelly,
Assistant Director
Address
One Gustave L.
Levy Place,
New York, NY 10029
Phone
212-241-6605
■

■
Albany Medical
Center
Contact
Claude Rounds,
Office Plant Manager
Address
43 New Scotland Ave.,
Albany, NY 12208
Phone
518-445-3243
■

Source Reduction in Other Hospitals

Other hospitals throughout the country have taken other source reduction initiatives. Some of them are described below.

Itasca County Medical Center in Minnesota conducted a source reduction pilot program at a 108-bed community hospital with an attached 35-bed convalescent nursing care facility. As a result, the hospital is preventing 238 cubic yards (over 10,700 pounds) of waste per year. (This is the only hospital case study found by INFORM that measured source reduction in both volume and weight.) These actions are saving $11,030 per year in purchasing costs alone, not including savings from avoided disposal fees.

Among the measures implemented are switching to reusables from disposables in food service (both in hospital cafeteria and patient rooms), changing to rechargeable batteries for nurses' flashlights, changing to longer lasting lightbulbs, and switching from milk cartons to pouches in the cafeteria. The switch to a bulk milk dispenser for patients alone keeps almost 75,000 milk cartons out of the waste stream each year.[13]

Mount Sinai Hospital in New York City decreased trash volume by 25 percent from source reduction measures. The hospital used to fill up eleven 35-cubic yard dumpsters each week, and now only uses eight. The reduction came from a switch from disposables to reusables in the cafeteria and from a switch from disposable to reusable bed pads in one-third of the beds (370 beds). Savings from the bed pads are estimated to be $56,000 in annual purchasing costs and $7000 in annual disposal costs. Approximately 200 tons of waste were reduced from this measure in 1991.[14]

Albany Medical Center in Albany, New York, reduced its waste stream by 425,000 pounds and saved approximately $93,500 in avoided disposal costs in 1991 by implementing a variety of source reduction measures. The hospital increased the use of reusables in the cafeteria, changed to reusable medical waste containers, composts kitchen and food waste, and uses cloth diapers in some units. In addition, it switched to bed pans made of recyclable material, a practice that is not source reduction but that is included in the hospital's waste and cost savings figures. Cost savings are high because most of this waste was regulated medical waste (RMW) and disposal costs of RMW are more than $400 per ton for Albany Medical Center.[15]

Northern Dutchess Hospital in Dutchess County, New York, conducted a cost analysis of cloth versus disposable diapers that projected the hospital could save approximately $1400 a year by switching to reusables. The switch would also reduce garbage by over 4500 pounds a year.[16]

The seven hospitals in Seattle, Washington, that deliver babies have returned to using cloth diapers from disposables. The change was made after the King County Nursing Association published its findings on the economic, environmental, and health impacts of disposable versus cloth diapers.[17]

The University of Mississippi at Jackson reports that costs of disposable medical instruments used by surgeons are far greater than the conventional reusable instruments. A set of disposables costs about $800, whereas reusable instruments cost $1500 for a set and can be used up to 20 times. Thus, for every set of disposables that is replaced by reusables, the hospital saves $14,500 in purchasing costs. In contrast, the makers of disposable instruments claimed, in a *New York Times* article, that their products are safer and more reliable and that sterilizing, repairing, and maintaining reusable equipment can be costly; however, INFORM could not compare costs based on the information provided in the article.[18]

Baptist Medical Centers in Birmingham, Alabama, reported that one hospital using reusable pillows saved $41,600 in purchasing costs compared to a hospital using disposable pillows.[19] Cleaning costs were not calculated.

❖ Schools

Schools are an effective place to begin source reduction programs for two reasons. First, reaching children at an early age is a good way to encourage positive environmental habits. And second, substantial reductions in the waste stream can be made by reducing waste generated by the approximately 60 million students enrolled in kindergarten through college[20] and school faculty and staff. Individual students generate an estimated 0.08 to 0.12 tons of waste per year,[21] or 4.8 million to 7.2 million tons of waste nationwide — 2.6 to 4 percent of the US waste stream.

Currently, garbage management education efforts in schools focus on recycling, but some schools are beginning to adopt source reduction practices. Examples of schools in the New York metropolitan area putting source reduction into practice include the Chappaqua Central School District in Westchester County and Rosyln High School on Long Island.

Chappaqua Central School District

■
Chappaqua Central
School District
Contact
Henne Starace,
Business Admini-
strator
Address
P.O. Box 21,
Chappaqua, NY
10514
Phone
914-238-7222
■

The Chappaqua Central School District returned to reusable trays from styrofoam disposables. The change was prompted by a committee formed by various school board members in the early 1980s. The committee received support from students in one elementary school (Westorchard School) who protested the use of styrofoam in the cafeteria by bringing plates in from home and washing them in the bathrooms. Cafeterias in this school district have now repaired old dishwashers and installed new ones.[22]

The Chappaqua Central School District business administrator reported that $44,000 was spent on new dishwashers and trays, and that the payback period is expected to be within two years. Minimal additional expenses are expected from the labor for dishwashing since current lunchroom custodians are taking over the responsibility. Preliminary estimates of the switch to reusables show a 50 percent reduction in the volume of trash being discarded from school cafeterias.[23]

■
Roslyn High School
Contact
James Carter, English
teacher, or John
Didden, Assistant
Principal
Address
Roslyn Road,
Roslyn, NY 11577
Phone
516-625-6386 or
516-625-6341

■

Roslyn High School

Rosyln High School on Long Island has an extracurricular club called Students Protecting the Environment Against Contamination (SPEAC) that was formed to make the school more environmentally conscious.[24] As part of this club, the students promote source reduction activities. The school has a strong recycling program and is just beginning to look at source reduction.

To date, most efforts have been geared toward encouraging both students and teachers to reduce paper use by posting notices for meetings, using smaller sheets of paper for short quizzes, sharing newspapers, and using paper already used on one side to write draft assignments. The program has support from the students who are encouraging teachers within the school to adopt these practices.

SPEAC also hopes to have the cafeteria return to reusable trays and tableware. The school still has trays and reusable tableware in storage, which could lower expected costs from returning to this system. SPEAC students are researching the economics of the return to reusables with appropriate administrators.

Other Source Reduction Possibilities for Schools

INFORM has identified several other source reduction measures that could be adopted within school systems.

- Placing swap boxes in schools or designating swap days for students to exchange items from home that are no longer used (such as outgrown clothing and sports equipment).

- Collecting unused school materials (such as notebooks, pens, and pencils) at the end of the year to donate to other students at the beginning of the next year.

- Making double-sided copies for handouts.

- Making note pads out of paper that has been used on one side.

Notes

1. CalRecovery, Inc. (Marian R. Chertow), *Waste Prevention in New York City: Analysis and Strategy,* prepared for the New York City Department of Sanitation, January 15, 1992, p. 21.

2. Personal communication, James Marion, Recycling Program Manager, New York State Department of Correctional Services, with Caroline Gelb, INFORM, November 25, 1991.

3. Information for this section was provided by James Marion, Recycling Program Manager, New York State Department of Correctional Services, in personal communications with Caroline Gelb, INFORM, November 25, 1991.

4. This figure is slightly overstated since it includes the weight of the outer shelling of the mattress which is not composted.

5. Tellus Institute (John Stutz and Gary Prince), *A Statistical Profile of New York City for Solid Waste Management Planning,* prepared for the New York City Department of Sanitation, May 17, 1991, pp. 3-26.

6. Waste-Tech, *The New York City Medical Waste Management Study, Task 4 Final Report: The New York City Medical Waste Management Plan,* June 24, 1991, section I, p. 2.

7. *Ibid.,* section II, p. 2.

8. *Ibid.,* section II, p. 5.

9. *Ibid.,* section II, p. 14.

10. Information for this section was provided by Dexter Dugan, Senior Management Consultant, New York City Health and Hospitals Corporation, in a letter to Bette Fishbein, INFORM, March 13, 1992.

11. New York City Department of Sanitation, *A Comprehensive Solid Waste Management Plan for New York City and Final Generic Environmental Impact Statement,* August 1992, p. 7-13.

12. Information for this section was provided by Cindy von Beren, Assistant to the President, Hospital of Saint Raphael, New Haven, Connecticut, in personal communication with Caroline Gelb, INFORM, March 4, 1991.

13. Minnesota Office of Waste Management, *Waste Source Reduction: A Hospital Case Study,* 1992, p. 1.

14. Personal communication, Michael Connelly, Assistant Director, Mount Sinai Hospital, to Caroline Gelb, INFORM, December 1991.

15. "Cure Waste" speech presented by Claude D. Rounds, Albany Medical Center, at the New York State Department of Environmental Conservation Third Annual Recycling Conference, November 13, 1991.

16. Based on information from Cornell Cooperative Extension of Dutchess County, November 19, 1991.

17. "Bottom Care," *In Business,* summer 1991, p. 40.

18. Milt Freudenheim, "The Tiniest, Kindest Cut of All," *The New York Times,* July 10, 1991, p. D1.

19. Phyllis Grant, Susan Laird, and Doris Coppage, "Disposable Versus Reusable Pillows: A Case Study," *Hospital Material Management Quarterly,* Vol. II, Issue 3, February 1990.

20. United States Census Bureau, 1990.

21. Tellus Institute, *op. cit.* Tellus estimates the waste generation factor per student to be 0.08 tons per year; the Washington Department of Ecology, 0.12; and RW Beck (a Colorado environmental consulting group), 0.09.

22. Patti L. Reid, "Students Push Cafeterias To Recycle and Re-Use," *The New York Times,* September 15, 1991, Westchester Weekly Section, pp. 1, 21-22.

23. Personal communication: Henne Starace, Business Administrator, Chappaqua Central School District, to Caroline Gelb, INFORM, January 1992.

24. Information for this section was provided by James Carter, English teacher and advisor of SPEAC, in personal communication with Caroline Gelb, INFORM, October 21, 1991.

▦ Chapter 8 Government Programs to Stimulate Source Reduction

G overnment can play an important role in motivating the private sector (businesses and residents) and helping it develop programs to reduce waste.
Strategies to achieve this include technical assistance and assistance in conducting
waste audits; assistance to businesses and residents for backyard composting,
grants, pilot programs, clearinghouses, awards and contests; and sponsorship of
reuse programs.

❖ Technical Assistance, Waste Audits, and Materials Assessments

Technical assistance programs are designed first to help businesses recognize
opportunities for waste reduction measures, and then to implement them. In
addition, government publicity about successful money-saving source reduction
programs can encourage innovation by other businesses.

The first step in any business source reduction program is conducting a waste
audit and a materials assessment. The waste audit identifies the materials that end
up in the trash can or recycling bin. The materials assessment identifies the
supplies, food, and other materials purchased by the company and its employees
that are brought into its facilities. This allows an analysis to be done of what
materials can be eliminated, reduced, replaced, and reused as well as recycled. It
also allows companies to identify which materials may end up in the trash of other
companies or consumers.

Some businesses are readily equipped to design their own tools for conducting
waste audits and materials assessments. Government may help others that are less
well equipped by designing worksheets, holding workshops, and giving assistance
to individual businesses.

The goal of a materials/waste audit for source reduction planning is to gather
information on the following:

- materials (supplies, equipment, food, etc.) purchased by the business
- packaging in which items bought by the company arrive
- materials brought into the office by employees
- documents received in the mail
- products and the packaging used to ship them from the company
- other materials leaving the company, such as reports, correspondence, faxes, and direct mail
- materials kept in the office (for example, in the supply room and the filing cabinets) that may be thrown away later
- materials that become waste
- materials separated for recycling

Once this information is gathered, source reduction programs can be developed to reduce specific materials.

In surveying government programs around the country, INFORM did not find any examples of audits that were specifically designed by government entities to encourage source reduction. Most were designed for assessing recycling opportunities. However, INFORM found examples of two technical assistance programs in the country specifically designed to assist businesses in reducing waste, although extensive waste audits or materials assessments are not included: WasteCap in Maine, Vermont, and New Hampshire; and a local program in Tompkins County, New York.

WasteCap in Maine, Vermont, and New Hampshire

WasteCap is a tristate (Maine, Vermont, and New Hampshire) public/private cooperative program structured to allow businesses to share hands-on source reduction and recycling information with each other. The focus is on solid waste, but there is also a specific provision for sharing hazardous waste information. The program relies on volunteers from the business community visiting other businesses and assessing waste reduction opportunities. The volunteers include engineers, recycling coordinators, brokers, and consultants.

In the Maine program, WasteCap serves as a catalyst for encouraging businesses to evaluate their operations from a waste perspective. The volunteers emphasize to clients that source reduction is the most effective waste strategy, both economically and environmentally, and is, therefore, good business.[1] Volunteers do not perform in-depth waste audits and, although both recycling and reuse opportunities are identified, source reduction receives the most emphasis. While WasteCap volunteers provide recommendations, it is up to individual companies to develop and implement changes in their waste management plans. WasteCap in Maine also does follow-up surveys with participating companies 6 and 12 months after the site visit to document the benefits of the program and quantify the overall achievements in waste reduction.

The Vermont program plans to start publicizing its projects in 1993 in state newspapers.[2] Since confidentiality clauses are included in contracts with participating businesses, only general, successful source reduction initiatives will be described unless companies give permission to use their names.

Tompkins County, New York

Tompkins County, New York, started a volunteer technical assistance program in the fall of 1991. Upon request, staff from the county's solid waste division provide free waste assessments to businesses. The assessment consists of a questionnaire, followed by a walk-through of the site during which county staff point out opportunities to reduce waste. Offices, stores, restaurants, gas stations, and schools have all taken part in the program.

■
Tompkins County
Solid Waste Division
Contact
Lynn Leopold
Address
Botswick Road,
Ithaca, NY 14850
Phone
607-273-6632
■

❖ Backyard Composting and Leave-on-Lawn Programs

Yard waste, which consists of leaves, branches, and grass clippings, is the second largest material category in the waste stream — 17.6 percent in 1988 (after paper, 40 percent) — according to the US Environmental Protection Agency. Food waste comprises 7.4 percent of the waste stream.[3] Yard and some food wastes present some of the greatest source reduction opportunities through backyard composting and leaving grass clippings on the lawn or mulching them. In addition to reducing waste, keeping these materials out of the waste stream reduces pollution from disposal systems since yard and food wastes create oxides of nitrogen when burned, and methane gas, leachate, and settling problems when landfilled.

Approximately 16 states and many cities throughout the country have banned yard wastes from incinerators or landfills. The choice then becomes municipal composting or source reduction. Municipal composting involves collection and processing at public expense and, therefore, is a recycling strategy. Municipalities in many localities have also had to deal with complaints about odors from composting sites from neighbors of these facilities. Municipalities that choose to promote backyard composting, by contrast, are promoting source reduction since the materials are handled on-site and never enter the waste stream. Backyard composting and leaving or mulching grass clippings on the lawn are the most efficient way to manage yard waste. In addition, the compost and mulched materials applied to yards and gardens improve soil quality and plant growth.

Two communities now successfully promoting reductions in yard waste are Seattle and New York City.

Backyard Composting in Seattle, Washington

The Seattle Solid Waste Utility's backyard composting program comprises the

■
Seattle Engineering
Department, Solid
Waste Utility
Contact
Carl Woestwin
Address
710 Second Ave.,
Suite 505,
Seattle, WA
981104-1709
Phone
206-684-4684
■

largest portion of the city's annual source reduction budget: $595,000 out of $871,000 in 1992 (as discussed in Chapter 4).[4]

Seattle is promoting yard waste reduction by giving residents free composting bins and having volunteers teach residents how to compost. Trained volunteers conduct community outreach for neighborhoods, business groups, schools, and the public at street fairs, festivals, and other citywide events. In addition, the city sponsors four composting demonstration sites throughout the city. In 1989, the city hired six full-time trainers who, by mid-1992, had given out 20,000 bins and completed thousands of in-home instruction sessions on composting techniques. Seattle also has a municipal composting program, but residents must pay to participate in it. For a fee of $3 per month, as many as 20 cans, bags, or bundles of yard waste will be collected.[5]

Seattle has used two surveys to measure the yard waste reduction from backyard composting. One was a survey of 275 bin recipients who were asked how many bins were filled and what percentage of waste they composted on site. According to their responses, the estimated average diversion from the municipal solid waste stream is in the range of 367-587 pounds per household per year. The other survey involved 31 volunteer households that weighed everything put in their bins for one year. This resulted in an estimated diversion of 952 pounds of yard waste per household per year, but the self-selected group was not considered representative of the population.[6]

Seattle's program has been successful financially. The Solid Waste Utility estimates savings from backyard composting to be $17.75 for each ton of yard waste not generated since the cost estimate for the backyard composting program is $68.41 a ton and the avoided disposal cost is $86.16 a ton.

Preliminary results of the Seattle Solid Waste Utility 1990 Waste Reduction Survey show that "most composters use informal composting methods such as mulching with yard waste around their plants (80 percent), or composting in piles (69 percent). A smaller proportion composts in bins (38 percent)."[7] Those who received home training composted more than those who did not.

In June 1992, Seattle began a food waste composting pilot study, involving 250 households randomly selected from those that received yard waste composters. They will weigh and compost food wastes for 6 months.

New York City Parks

Grass clippings can be left on the lawn to act as a fertilizer and prevent weeds. Every standard 30-gallon garbage bag of grass clippings contains up to one-quarter of a pound of organic nitrogen which can be used to enrich lawns. Leaving grass clippings on lawns also saves the time that would otherwise be required for bagging the grass and changing the bag on the mower.

The New York City Department of Parks and Recreation is required by a 1989 local law to compost its yard waste.[8] Prior to the passage of the law, most of the

■
New York City
Department of Parks
and Recreation
Contact
Kristen Harvey
Address
1234 Fifth Ave.,
Room 237,
New York, NY 10029
Phone
212-360-1406
■

organic materials (leaves, forestry debris, and grass) were collected by the Parks Department and taken to the landfill; the law will result in diverting this yard waste from the city's landfill. Any earlier composting programs were run by individual park workers on a small scale.

In response to the local law's mandate, the Parks Department has set up several on-site composting areas throughout the city. The agency is also building a large center for chipping park forestry debris (tree limbs, logs, and stumps). An informal policy has developed through which grass clippings and leaves are not collected and brought to composting sites unless absolutely necessary. Grass clippings are left on park lawns, and leaves are used as mulch in shrub beds and wooded areas where they will degrade naturally.

The compost and wood chips produced by the agency's composting program are being used in horticulture and maintenance operations throughout the city. In the future, some of this material will be made available to the public. A small amount of wood debris is being split into firewood and sold at local farmers' markets.

The Parks Department has not yet estimated total cost savings for the source reduction measures. Expected cost savings would be from avoided costs in purchasing bags and in collection costs. (No savings for disposal are involved because the city's Department of Sanitation does not charge other city agencies, including the Parks Department, for disposal.)

Leave-on-Lawn Programs and Other Initiatives

■
Don't Bag It,
Cornell Cooperative
Extension,
Nassau County
Address
1425 Old Country Rd.,
Plainview, NY
11803-5015
Phone
516-454-0900
■

Numerous communities have begun leave-on-lawn programs. In one, the New York State Department of Economic Development's Office of Recycling Market Development and the Nassau County (New York) office of the Cornell Cooperative Extension are promoting a "Don't Bag It" program in Nassau County, Long Island. As part of a promotional effort, a Cooperative Extension representative gave a speech about the benefits of leaving grass clippings on the lawn to several thousand landscapers at a trade show at the Nassau Coliseum.[9]

Another community, Hamilton County, Wyoming, began a "Don't Bag It" campaign in March 1991. Residents who do not compost or leave grass clippings on the lawn must pay $0.50 a bag to have their yard waste collected. Although this fee is small, the county hopes it will encourage residents to manage waste on their property.[10]

Yard and food waste source reduction activities are also beginning on the federal level. In 1991, Representative George Hochbrueckner (D-NY) proposed a bill (HR 2138) that would establish a pilot program for composting of yard and cafeteria waste at a Department of Defense facility.[11] As of mid-1992, the House Armed Services Committee had not acted on the bill. However, some 20 military bases have begun composting programs that go beyond the original pilot program idea. The bill may be redrafted to make such programs mandatory on all bases.

❖ Grants

Government can give grants to industry or community groups to develop innovative solutions to reducing waste. Grants can be administered on the state or local level. Three grant programs that include allocations of funds specifically for source reduction projects are run by the Michigan Department of Natural Resources, the Vermont Agency of Natural Resources, and Seattle's Solid Waste Utility.

Michigan's Solid Waste Alternatives Program

■
Michigan Department
of Natural Resources,
Waste Management
Division
Contact
Sharon L. Edgar,
Unit Chief, Solid
Waste Alternatives
Program
Address
PO Box 30241,
Lansing, MI 48909
Phone
517-373-4749
■

In 1988, Michigan voters approved a bond act that would fund solid waste projects, and the state legislature approved $150 million for the "Protecting Michigan's Future Bond Solid Waste Alternatives Program" (SWAP).[12] Located within the Michigan Department of Natural Resources, SWAP began operating in fiscal year 1989. As of early 1992, it had received a total of 860 applications for projects in thirteen different funding categories, of which 281 were approved. These approved projects involved almost $97 million in grant and loan funds.

One of the funding categories is Waste Reduction Research and Demonstration (WRRD), in which over $684,000 has been approved for seven projects. While this is less than one percent of SWAP's total grant money, there had been only 18 applicants in this category by early 1992. The Department of Natural Resources is making an effort to make more Michigan companies aware of the WRRD category under SWAP, in hopes of increasing the number of high-quality proposals that focus on source reduction.

Two of the seven source reduction grants and loans allocated by SWAP are a loan to a Tiny Tot Cloth Diaper Service to increase its capacity so it can serve more households, and a grant to Michigan State University Department of Horticulture to test, in 200 households, strategies that could reduce solid waste generation by 30 percent. Since the latter grant was given in the fiscal year 1991-92, results were not available when this report went into production.

Vermont Waste Reduction Incentive Grants

■
Vermont Agency of
Natural Resources
Contact
Paul Markowitz
Address
103 South Main St.,
Waterbury, VT 05676
Phone
802-244-7831
■

The budget for the Vermont Agency of Natural Resources Waste Reduction Incentive Grant Program was $138,000 in 1990; however, some of the grants have been allocated to projects to reduce materials such as hazardous waste and concrete that are not part of the municipal solid waste stream. Two of the source reduction grants are $25,000 to Associated Industries of Vermont to implement WasteCap (see discussion of WasteCap earlier in this chapter) and $11,000 to Medical Center Hospital of Vermont to study and implement a source reduction and recycling program in the surgical service wings, and to distribute a state government manual about source reduction and recycling to all Vermont hospitals.[13]

■
Seattle Engineering
Department, Solid
Waste Utility
Contact
Carl Woestwin
Address
710 Second Avenue,
Suite 505,
Seattle, WA
98104-1709
Phone
206-684-4684
■

Seattle Solid Waste Utility Grant Program

Seattle's Solid Waste Utility's source reduction budget allocated $80,000 (out of an $800,000 budget) to its grant program in 1991.[14] The funded projects, listed below, received grants ranging from $2500 to $19,000.

- Puget Consumer Co-op, which has six stores and 37,000 members, purchased reusable plastic containers to replace waxed corrugated boxes used for produce.

- King County Nurses Association conducted education activities on alternatives to "adult diapers."

- The Seattle school district received money to purchase 120 worm bins so classrooms can do food composting.

- Two community groups organized rummage sales at which source reduction was encouraged.

- A day care center is exploring various source reduction options.

- Volunteers from the Utility's Friends of Recycling program set up and give tours of two model waste reduction homes.

- Washington Citizens For Recycling will work in community service organizations to support source reduction.

❖ Pilot Programs

Government-sponsored pilot programs can provide an opportunity to test and demonstrate source reduction strategies. The information and data generated can be valuable for expanding and improving these pilots into comprehensive municipal programs. Two ambitious pilot programs are run by the Minnesota Office of Waste Reduction in Itasca County, and the New York City Department of Sanitation in two communities (Park Slope in Brooklyn, and the Upper East Side of Manhattan).

■
Minnesota Office of
Waste Management
Contact
Kenneth Brown,
Waste Reduction
Management
Address
1350 Energy Lane,
St. Paul, MN 55108
Phone
612-649-5743
800-652-9747
(toll free in MN)
■

Itasca County, Minnesota

The Minnesota Office of Waste Management is conducting source reduction demonstration projects in Itasca County. Reports on three projects have been issued: one at the county courthouse and the city garages; one at Itasca's Grand Rapids Herald Tribune, a newspaper; and one at Itasca Medical Center (discussed in Chapter 7).

One project is being conducted within Itasca County's courthouse and 16 Road & Bridge Department garages. As a result, the county reduced approximately 10 percent of the waste generated by these facilities and is saving $42,000 and preventing 9500 pounds of waste per year. Among the measures implemented are:

- Changing to cloth roll towels from paper.
- Switching to stainless steel air filters in the furnace and air conditioning systems instead of disposables. This worked well in the garages, but did not work in the courthouse. The courthouse is now going to try using reusable static electric filters.
- Purchasing cleaning products in reusable 5-gallon pails at the garages.
- Purchasing long-lasting, repairable, and durable products, such as chainsaws and brush saws. (This measure was already in place.)
- Reducing direct mail and duplicate mail received by collecting it and periodically sending memos to the senders asking to be removed from the mailing lists.
- Increasing two-sided copying and using single-sided paper for scratch pads.
- Eliminating 275 gallons of equipment paint, which saves $16,500 annually.
- Using cloth rags at the garages (this did not work because the wastewater treatment system does not handle grease and oil well).[15]

Another project is being conducted at *The Grand Rapids Herald Tribune,* which publishes a biweekly newspaper with a circulation of 8000 and a weekly advertiser with a circulation of 20,000. The newspaper is saving $12,914 per year through changed procurement policies and operations, an additional $3900 per year from reducing the number of garbage dumpsters haulers must pick up, and an additional $1608 per year on its janitorial cleaning service since the decrease in garbage means that less cleaning is needed. This brings the total annual savings from source reduction measures to $18,422, without considering avoided disposal costs. Waste going to the county landfill was reduced by 25,150 pounds per year.[16]

Among the strategies the newspaper has adopted are:

- Decreasing overruns of the publications from 250 to 50 copies per issue, which saves on paper, ink, and labor.

  Weight of waste avoided: 17,000 pounds per year

  Cost savings: $7300 per year

- Selling the leftover paper on the newspaper stock spools to a ceramic packaging firm that packs with it instead of foam.

  Weight of waste avoided: 7537 pounds per year

  Cost savings: $1809 per year from sales

- Collecting and reusing excess colored and black ink, which saves on the cost of virgin materials and decreases the amount of ink containers discarded. (The paper is considering switching to soy-based black ink in order to reduce toxics, but soy-based colored ink is considered too expensive at this point.)

Weight of waste avoided: 2100 pounds per year

Cost savings: $2615 per year

- Reducing the amount of chemicals needed in film developing by not throwing away weekly leftovers. The newspaper went from using 1 gallon a week to 1 gallon every 3 weeks. This results in both toxics and quantity reductions.

Weight of waste avoided: 210 pounds per year

Cost savings: $140 per year

- Rolling and loading only the amount of film needed by each photographer each day to reduce the number of unshot frames.

Cost savings: $60 per year

- Switching from wide-ruled to narrow-ruled reporters' notebooks.

Weight of waste avoided: 34 pounds per year

Cost savings: $96 per year

- Reusing paste-up (layout) sheets by peeling off the layouts. Sheets are now reused an average of six times instead of being discarded after a single use.

Weight of waste avoided: 285 pounds per year

Cost savings: $570 per year

- Switching to reusable toner cartridges for computers and photocopiers.

Weight of waste avoided: 27 pounds per year

Cost savings: $900 per year

- Reusing single-sided copy paper for note pads and for labels instead of buying new labels.

Weight of waste avoided: 120 pounds per year

Cost savings: $180 per year

- Changing to reusable cloth towels from disposable paper towels.

Weight of waste avoided: 135 pounds per year

Cost savings: $120 per year

- Keeping labels from partially used label sheets used for mailings.

Cost savings: $20 per year

- Selling aluminum printing plates for reuse. (Any not sold for reuse are recycled.)

Weight of waste avoided: 4250 pounds per year

Cost savings: $1200 per year from sales

The Minnesota Office of Waste Management source reduction coordinator publicized this case study by speaking at the Minnesota Newspaper Association conference. The Association also ran a series of articles about the source reduction program in its monthly bulletin. Publicity about this project may encourage other larger newspapers to adopt similar measures.

■
New York City Department of Sanitation, Unit of Waste Prevention and Related Projects
Contact
David Kleckner
Address
44 Beaver St.,
New York, NY 10004
Phone
212-837-8177
■

New York City

New York City's pilot programs fall into two categories: a program for residential communities and a cooperative program with businesses. In the spring of 1990, the New York City Department of Sanitation (DOS) organized 13 buildings (with 3000 residents) on the Upper East Side of Manhattan and in East Harlem to participate in a small-scale intensive waste prevention and recycling pilot project, called a mini-pilot. (New York City uses the term "waste prevention" to mean source reduction.) The mini-pilot project was designed to test certain source reduction strategies as a precursor to a larger pilot project. The participating residents came from a cross section of incomes in a high-density housing area. Among the DOS initiatives were:

- Distributing reusable canvas and string shopping bags to stores.

- Mailing brochures with tips for reducing waste at home, work, and stores to residents.

- Sending information to individuals about how to reduce specific materials that become waste, such as disposable diapers and direct mail.

- Giving residents directories, tailored for individual buildings, that listed local repair shops, thrift shops, and charity groups that accepted donations of used household items.

DOS was not funded to do follow-up evaluations to see if participating residents had changed their behavior to reduce the amount of waste they generated. Although there was no formal follow-up work, DOS observed that few people actually used the string bags for shopping and that some residents thought the "Householder's Guide to Waste Prevention" brochure was too obvious. DOS has since revised the household tips pamphlet. Participating residents thought the list of repair and charity organizations was very useful.[17]

The full-scale pilot program that developed out of the mini-pilot is the "Intensive Zone," which comprises two sections of Community District 6 (24,000 households) in Park Slope, Brooklyn. It encompasses an area of medium residential density and medium income that was already recycling newspaper, magazines, corrugated cardboard, metal and glass containers, and bulk items. The original design of the program was very comprehensive, but due to budget constraints many projects have been either put on hold or canceled. DOS chose to implement those components of the program that appeared to be most cost effective and to have the highest potential to reduce waste.

Components implemented:

- Backyard composting education and demonstration project (currently in place though there is no longer funding to properly maintain the demonstration site).

Components in planning stages:

- Outreach to residents: The revised "Householder's Guide to Waste Prevention," which explains how to reduce both the amount and toxicity of waste generated within households, will be distributed door to door in certain parts of the Intensive Zone by volunteers from the local recycling center. Additional guides will be available in central locations.

- Retail business outreach: A source reduction coordinator will meet with the local merchants' association to encourage businesses to hang posters asking customers to use bags only when needed.

Components on hold:

- School curriculums on waste prevention.

- Outreach to all private, nonprofit, and public offices and institutions to educate staff and management on how to reduce waste.

- Work with manufacturers to test market products that generate less waste (such as concentrates and refillable containers).

- Technical assistance and demonstration of waste prevention techniques for businesses and institutions to help them conduct waste audits, and general discussions with retailers on how to reduce waste.

Components cancelled due to lack of funds:

- A waste prevention advertising campaign in mass transit hubs of the zones. (The city already has waste prevention posters in subways.)

- A waste prevention video or slide show to enhance outreach to institutions and community groups.

- Contracts with existing organizations that sell or donate used items to expand programs in the Intensive Zone, and to enhance effectiveness throughout the city.

- Distribution of reusable cloth bags for shoppers to stores.[18]

In September 1991, DOS announced the creation of a citywide Waste Prevention Partnership. The department is working with businesses and business associations throughout the city to design and implement "waste prevention" measures. To date, a supermarket chain (D'Agostino's), the New York State Food Merchants Association, Inc., the Direct Marketing Association, and the Neighborhood Cleaners Association have joined the partnership.

D'Agostino's and the Food Merchants Association are encouraging customers to bring reusable grocery bags to stores and are using brochures to teach consumers how to be less wasteful shoppers. D'Agostino's has instituted a program of not providing double bags unless they are requested by a customer. As a result of this measure, expenditures on bags have been cut in half despite an increase in grocery sales. D'Agostino's is also planning to pay customers 2 cents for not taking a bag and bringing one of their own instead.

The Neighborhood Cleaners Association is promoting the return of hangers to dry cleaners and exploring the option of using returnable/reusable nylon garment bags. About 90 percent of the Association's 1200 members, or close to 1100 cleaners, are located in New York City, and 95 percent of the city's dry cleaners belong to the Association. Return of hangers and garment bags can save a cleaner $3000 per year in purchasing costs and reduce the waste generated when consumers dispose of these items.

The Direct Marketing Association is continuing its existing efforts to reduce direct mail's contribution to the waste stream. New York City sends 70,000 tons of unopened mail each year to Fresh Kills landfill.[19] (Direct mail is discussed in more detail in Chapter 10).

Dutchess County, New York

■
Dutchess County
Environmental
Management Council
Address
PO Box 259, Farm
and Home Center,
Millbrook, NY 12545
Phone
914-677-3488
■

Dutchess County, New York, lies some 100 miles north of New York City and includes both rural and urban neighborhoods. The county's Environmental Management Council is beginning a one-year source reduction pilot program, developed with INFORM's assistance, in the fall of 1992. The program will be staffed by one full-time source reduction specialist. The specialist will first educate targeted waste-generating sectors (hospitals, schools, small businesses, consumers, and county and local government agencies) about the need for source reduction. Then the specialist will work with waste generators in these sectors to help them identify and implement specific source reduction measures. The effects of the pilot program will be measured and evaluated at the end of one year, and the findings will be publicized to encourage other institutions, businesses, and offices to adopt similar source reduction strategies.

❖ Clearinghouses

Through information clearinghouses, government can share source reduction program ideas with community activists, businesses, residents, and other government officials. The Minnesota Office of Waste Management (OWM) is one example of a government agency that has created a clearinghouse that is beginning to focus on source reduction.

Minnesota

The Minnesota legislature created the Waste Education Program in 1986 to provide a unified approach to waste education throughout the state. The objective

■

Minnesota Office of
Waste Management,
Waste Education
Program
Contact
Linda Countryman or
Nancy Skuta
Address
1350 Energy Lane,
St. Paul, MN 55108
Phone
612-649-5750
800-652-9747
(toll free in MN)

■

of the program is to increase the awareness of waste management methods that have less harmful impacts on the environment and help business, industry, local governments, and the public make more informed decisions on waste management issues. As part of increasing public awareness, the Waste Education Program created the Waste Education Clearinghouse.

The clearinghouse distributes information on subjects that include source reduction, recycling, composting, incineration, landfills, household hazardous waste, and other hazardous waste. Most requests for information come from teachers, but some also come from citizens, business representatives, and solid waste officials. Most of the average of 150 requests received each week refer to recycling issues. In order to promote source reduction, clearinghouse staffers tell callers about its importance and ask them if they would also like related source reduction materials. The clearinghouse provides fact sheets on source reduction and case studies of pilot programs across the state. It also has an expanding collection of source reduction materials and educational videotapes.

The philosophy of the Waste Education Program has been to promote and distribute preexisting materials. However, when the program identifies areas that could use additional information, the program has responded by developing its own materials, such as a *Community Waste Education Manual* that focuses on how to develop a public waste education campaign. The manual was designed for local governments and citizens' groups to use in public awareness campaigns. OWM is considering writing a supplement to the manual to make it more comprehensive, which would include providing details on source reduction and backyard composting.

❖ Awards and Contests

Governments can sponsor contests and give awards to encourage businesses, institutions, and manufacturers to adopt source reduction practices and to increase consumer awareness about source reduction opportunities. While there are many environmental contests and awards, most focus on recycling. The US Environmental Protection Agency presents annual Administrator's Awards for a variety of environmental achievements and the Council on Environmental Quality (within the Executive Office of the President) gives out annual environmental awards under its President's Environmental and Conservation Challenge Awards program. Even though there is no category for source reduction, some companies have received awards for reducing waste generation. The Rhode Island Solid Waste Management Corporation planned to conduct a packaging design competition, but it was never funded.

Private companies are also beginning to use awards to promote source reduction within their operations. For example, Ace Hardware, a retail chain, has developed a program that gives awards to hardware manufacturers accomplishing source reduction.

US Environmental Protection Agency Administrator's Awards Program

The United States Environmental Protection Agency established an annual Administrator's Award Program in 1991 to recognize environmental accomplishments by businesses, governments, environmental and community groups, non-profit and trade organizations, and educational institutions. While the program does not focus on municipal solid waste source reduction (the 1992 awards were given for pollution prevention projects), it provides a model for high-level public recognition of environmental achievements.

One of the 1992 winners, Chrysler's Jefferson North Assembly Plant in Detroit, received its award for the overall pollution prevention strategies built into the design of the plant. The award also cited one specific solid waste source reduction accomplishment: the plant's "no landfill goal." Some 65 percent of all parts are now shipped to Jefferson North in returnable containers.

President's Environmental and Conservation Challenge Awards

The President's Environmental and Conservation Challenge Awards include four categories of awards: (1) partnership (companies that take a cooperative approach with other organizations to tackling environmental problems), (2) environmental quality management (companies that use environmental practices themselves), (3) innovation (companies that develop creative technologies that have less harmful environmental impacts), and (4) education and communication (programs that raise the public's environmental awareness). In 1991, one company received an award and four received citations for source reduction initiatives, out of a total of 9 awards and 23 citations given. The award winner was the Waste Reduction Task Force of McDonald's Corporation and the Environmental Defense Fund for developing and implementing waste reduction and recycling programs that they expect to reduce McDonald's waste stream by 80 percent.

Ace Hardware Corporation

■
Ace Hardware
Corporation
Contact
Linda Baechtold,
Advertising
Coordinator
Address
2200 Kensington
Court,
Oak Brook, IL 60521
Phone
708-990-2821
■

Ace Hardware Corporation, a dealer-owned cooperative, started an environmental awards program at its dealer convention in 1991. At this convention, about 950 vendors displayed new products that Ace dealers could sell in their 5300 retail stores. There are three categories of environmental awards: manufacturing awards for improvements in manufacturing processes that help the environment; product awards for product improvements or new products that reduce impacts on the environment; and packaging awards for the packaging improvements that reduce impacts on the environment.

Although the categories do not place emphasis on source reduction, two of the 1991 Environmental Award winners were companies that promote source reduction. In the packaging category, Pfalzgraff was a certificate winner (runner up) for safely reducing the amount of packaging used for highly breakable items. In the manufacturing category, Arrow Plastics was a certificate award winner for reducing the amount of waste it generated by 20 percent (from 3850 cubic yards a year to 3070 cubic yards a year).

Other Ideas for Awards and Contests

INFORM has identified several other project areas that governments could target for source reduction awards or contests:

- Innovative source reduction initiatives within different industries, such as retail, food establishments, car services, and offices.
- Product design changes that promote source reduction, such as increasing durability, repairability, and availability of replacement parts.
- Education campaign slogans.
- Development of businesses that promote source reduction, such as repair shops.

❖ Reuse Programs

Governments can sponsor, promote, and publicize programs that distribute used goods for reuse. These programs keep materials that would otherwise be discarded out of the waste stream and, thus, give items a second life. In addition, items are made available at lower costs.

The concept of reuse is not a new idea. Yard sales are a billion dollar a year industry.[20] Antique stores and used record and book stores are widespread. Charitable donation organizations, such as Goodwill Industries and The Salvation Army, handle millions of tons of clothing and other household items each year.[21]

In addition to sponsoring and promoting reuse programs, government can publicize independent (profit and nonprofit) programs. This can encourage others to donate goods, use these services, and start other reuse programs.

The three reuse programs profiled here include one example of a government-sponsored program (Materials for the Arts), and several examples of independent programs that government could promote or publicize. Each targets materials that are significant contributors to the waste stream. Materials for the Arts in New York reduces furniture and furnishings (4.2 percent of the waste stream) and paper, beads, and buttons (materials that are classified as nondurable goods, a category that overall makes up 28.1 percent of the waste stream). City Harvest in New York reduces food waste (7.4 percent of the waste stream). Goodwill Industries reduces furniture and furnishing waste as well as clothing and shoe wear waste (2.2 percent of the waste stream).[22] Another kind of reuse project, reclaiming latex paints, is discussed in Chapter 13 on reducing toxic materials in the waste stream.

■
Materials for the Arts
Contact
Susan Glass
Address
410 W. 16th St.,
New York, NY 10011
Phone
212-255-5924
■

Materials for the Arts, New York City

The New York City Department of Cultural Affairs and Department of Sanitation co-fund Materials for the Arts (MFA), a reuse program that matches donations from businesses and individuals with nonprofit arts organizations and public art projects. Materials, collected from approximately 1000 businesses and individuals, include furniture, paper, beads, buttons, and art/film supplies. Materials are picked up free and stored in a warehouse until distributed.[23]

MFA processes an average of 32 to 35 tons of materials a month. Since 1979, over $10 million in materials have been donated. Some examples of donations are: hundreds of gallons of paint from Donald Kaufman Color and A. Kraus and Sons that have been used for refurbishing offices, lobbies, stage sets, and murals; audiovisual equipment from Reeves Corporate Services, Film Equipment Rental Co., and Meeting Makers that has enabled film/video and performance groups to expand services and improve facilities; and musical instruments and performance equipment from Radio City Music Hall and Madison Square Garden.

MFA has also co-sponsored "Materials for the Schools" with the Board of Education. Representatives from over 400 schools received donated materials from businesses.

■
City Harvest
Contact
John Mooney
Address
159 W. 25th St.,
New York, NY 10001
Phone
212-463-0456
■

City Harvest, New York City

Governments can help publicize reuse programs such as City Harvest, a food distribution program in New York City. Distributing edible surplus food reduces waste disposal costs and contributes to the reduction of the waste stream, while benefitting thousands who might otherwise go hungry. In addition, private food distribution can help reduce social service costs for governments and nonprofit organizations.

City Harvest was established in 1982 to collect food that would otherwise go to waste and donate it to organizations such as homeless shelters, day care centers, and senior citizens programs. Food is donated by restaurants, corporate cafeterias, retail stores, and individuals on a daily basis.

City Harvest delivers an average of more than 10,000 pounds of food per day. Deliveries totaled 3.2 million pounds in 1989, 4 million pounds in 1990, and 4.6 million pounds in 1991. Total operating expenses in 1991 were $1.5 million, over 50 percent of which was for salaries. City Harvest is funded by individuals, foundations, and corporations.[24]

Goodwill Industries

Governments can also fund or promote the development of organizations that receive materials (such as clothing and furniture) from individuals and sell the materials at discount prices to the public. Redistribution of these materials reduces waste. One such organization is Goodwill Industries, which uses revenue from the

■
Goodwill Industries of
America
Contact
Bill Rable, Director of
Communications
Address
9200 Wisconsin Ave.,
Bethesda, MD 20814
Phone
301-530-6500
■

■
Goodwill Industries of
Greater New York, Inc.
Contact
Percy Preston, Jr.,
Director of
Development
Address
4-21 27th Ave.,
Astoria, NY 11102
Phone
718-728-5400
■

sale of used materials and other donated goods to provide job training to people who are disabled or have other barriers to employment, such as homelessness or illiteracy. One benefit to government, in addition to the reduction in the waste stream, is that Goodwill Industries employs prior government assistance recipients. In 1990, people trained by Goodwill Industries and either employed there or elsewhere paid $49 million in local, state, and federal taxes.[25]

In 1990, Goodwill stores throughout the United States sold 556.9 million pounds of clothing, in addition to goods such as furniture and automobiles. An estimated $325 million in sales from clothing and other items was generated in that year, up 9 percent from 1989.[26] Less than one percent of the materials given to Goodwill were thrown out. Materials that are not suitable for sale in the stores are sold to salvage companies to make industrial rags, sell overseas, or otherwise use.[27]

Goodwill Industries has now expanded its services at facilities throughout the country, further promoting source reduction. Two programs in particular, in Minneapolis and New York City, receive customer returns, damaged merchandise, and other items that would otherwise be thrown out from retail stores, and then sell the items in Goodwill stores. New York City's Goodwill (Goodwill Industries of Greater New York, Inc.) receives returned toys from the Toys R Us toy store chain, which are sold at lower prices in Goodwill stores in New York City and in Harrison, New Jersey.[28]

Goodwill Industries in Minneapolis receives materials from the Target store discount chain. The items are then clearly marked as Goodwill merchandise to prevent people from making returns to Target. An itemized list of donations is then returned to the chain for tax or documentation purposes. Sears and J C Penney are among the other national chains with which Goodwill Industries has worked. At present, Target is the only chain that has a national agreement with Goodwill Industries.[29]

Other Reuse Initiatives

The City of Baltimore, Maryland, recently began a test project to reduce waste. Residents in one neighborhood are asked to place reusable goods, such as clothing and appliances, at the curbs, where they are collected by Goodwill Industries and the Salvation Army.[30]

The San Francisco Recycling Program is promoting the reuse of live Christmas trees through a demonstration program at City Hall each holiday season. The trees have signs on them that say the same tree will be back next year and will be even bigger.[31]

Notes

1. Information for this section was provided by Jody Harris, Maine WasteCap Coordinator, in personal communications with Caroline Gelb, INFORM, June 1991.

2. Information for this section was provided by Connie Leach, Vermont WasteCap Coordinator, in personal communications with Caroline Gelb, INFORM, September 23, 1991.

3. US Environmental Protection Agency, *Characterization of Municipal Solid Waste in the United States: 1990 Update,* June 1990.

4. Information for this section was provided by Carl Woestwin, Seattle Solid Waste Utility, in personal communications with David Saphire, INFORM, September 9, 1991.

5. City of Seattle, *Seattle's Integrated Solid Waste Management Plan,* August 1989, pp. 43-47.

6. David Allaway, "Does Source Reduction Work?" *Resource Recycling,* July 1992, pp. 52-59.

7. Karen A. Brattesani, Ph.D., of Research Innovations, prepared for The Hartman Public Relations Group, *Seattle Solid Waste Utility 1990 Waste Reduction Survey,* April 1990, p. 30.

8. Information for this section was provided by Kristen Harvey, Organic Recycling Division, New York City Department of Parks and Recreation, in personal communications with Caroline Gelb, INFORM, October 29, 1991.

9. Margaret Roach, "Mow It, but 'Don't Bag It'," *New York Newsday,* March 21, 1991, p. 91.

10. *Hamilton County Solid Waste Management District News,* Volume 2, Issue 2, Second Quarter 1991, p. 1.

11. "Federal Watch," *Resource Recycling,* July 1991, p. 16.

12. Information for this section was provided by Sharon L. Edgar, Unit Chief, Solid Waste Alternatives Program, Waste Management Division, Michigan Department of Natural Resources, in a letter to Caroline Gelb, February 13, 1992.

13. According to statistics from the Vermont Agency of Natural Resources, November 1990.

14. Information for this section was provided by Carl Woestwin, Seattle Solid Waste Utility, in personal communications with David Saphire, INFORM, September 9, 1991.

15. Minnesota Office of Waste Management, *Waste Source Reduction: County Courthouse and Garage Case Study, May 1992 Update,* pp. 1-9.

16. Minnesota Office of Waste Management, *Waste Source Reduction: Small Business Case Study, 1991,* pp. 1-7.

17. Personal communication, Lisa Fernandez, Recycling Programs Planning Division, New York City Department of Sanitation, to Caroline Gelb, INFORM, June 1991.

18. New York City Department of Sanitation, *New York City Recycles: Preliminary Recycling Plan, Fiscal Year 1991,* pp. 24, 27, and 28.

19. New York City Department of Sanitation, Waste Prevention Partnership Press Kit, September 25, 1991.

20. Gretchen Mary Herrmann, "Garage Sales As Practice: Ideologies of Women, Work and Community in Daily Life" Volumes I and II, Dissertation: State University of New York at Binghamton, 1990, abstract.

21. Information for this section was provided by David Barringer, Director of Communications, Goodwill Industries of America, in personal communications with Caroline Gelb, INFORM, September 30, 1991.

22. Waste stream percentages from US Environmental Protection Agency, *Characterization of Municipal Solid Waste in the United States: 1990 Update,* June 1990.

23. Information for this section was provided by Susan Glass, Materials for the Arts, in personal communications with Caroline Gelb, INFORM, April 1991.

24. Audited financial statements by Ernst & Young, City Harvest, Inc., June 30, 1991, p. 3.

25. Goodwill Industries of America, Inc. "Facts," 1990.

26. Goodwill Industries of America, Inc. 1990 Annual Report.

27. Personal communication, David Barringer, Director of Communications, Goodwill Industries of America, to Caroline Gelb, INFORM, September 30, 1991.

28. Personal communication, Percy Preston, Jr., Director of Development, Goodwill Industries of Greater New York, Inc., to Caroline Gelb, INFORM, September 30, 1991.

29. "Re-Use a Good Sign for Goodwill," *Retailing & The Environment/Discount Store News,* March 18, 1991, p. 90.

30. "Programs in Action," *Resource Recycling,* July 1991, p. 28.

31. Amy Perlmutter, San Francisco Recycling Program Director, speech at the New York State Legislative Commission on Solid Waste Management's Eight Annual Conference on Solid Waste Management, January 31, 1991.

◼ Chapter 9 Publicizing Business Source Reduction Programs

With the commercial sector generating about 40 percent of this country's waste, there are many opportunities for businesses to reduce their wastes. Government can promote source reduction within the private sector by publicizing the financial and environmental achievements of businesses that have successfully implemented source reduction programs. In addition, government agencies could emulate initiatives adopted by businesses. Six groups of business source reduction programs found by INFORM are profiled below: AT&T, the airline industry, Allied-Signal's Bendix Automotive Systems, Herman Miller, Inc., the hotel industry, and Central Hudson Gas & Electric Company.

Many of these examples focus on reusable shipping materials and on procurement (which was discussed in Chapter 6). Ninety to ninety-five percent of all US manufactured consumer, wholesale, and industrial goods are shipped in disposable corrugated boxes.[1] Disposable containers are used because they are convenient, light in weight (which lowers shipping costs), and do not require storage (since they are immediately thrown out or recycled). Corrugated cardboard comprises 12.9 percent of the national waste stream; it is the largest product category.[2] Many companies are exploring ways to return to reusable shipping materials and reduce the amount of packaging used, thereby both reducing waste and saving money.

Administrators running the programs profiled below agree there are two main steps to beginning a successful source reduction program. First, support from the head of the company is necessary. And second, programs benefit from employee involvement in suggesting source reduction strategies, either through a waste reduction group or a suggestion box. Involving employees pulls people from different aspects of the business together, so waste reduction efforts can be discussed on many levels. It allows people in different areas, such as in the manufacturing processes, administrative offices, and computer division, to make suggestions for their own areas and, thus, find the greatest opportunities for reducing waste.

AT&T

AT&T has used procurement policy to improve the performance of photocopy machines. The company set a source reduction goal to decrease paper use by 15 percent by 1994 (from 1990 levels) and to increase double-sided copying (duplexing). At the first centralized photocopy unit to start using the plan, the duplex rate increased from 10 to 79 percent. For the company as a whole, AT&T estimates that annual paper savings, if double-sided copying is increased to 50 percent company-wide, will be 77 million sheets of paper, which will reduce purchasing costs by $385,000.[3]

AT&T has worked with its suppliers of copy machines to promote two-sided copying. AT&T's machines are being retrofitted to "make duplex the default mode." This means the machines will automatically copy two-sided unless specifically programmed to do one-sided copies. In addition, the copy machine manufacturers are working to assure that the recycled paper specified by AT&T procurement guidelines will not cause jams when machines are used in duplex mode.

Efforts by AT&T to improve duplexing performance of machines are likely to have benefits well beyond that specific company. Manufacturers, in order to preserve or increase market share, respond to customer concerns. If large buyers of copy machines, like AT&T, make double-sided copying a top priority, the copy industry is likely to respond and improve the duplexing performance of its machines. Other office product companies may follow suit. Hewlett Packard, for example, has developed a laser printer that makes two-sided copies.

Airlines

Although there are no statistics on how much waste is generated by airlines, the Federal Aviation Administration estimates that approximately 20,000 flights depart each day from US airports with an average of 70 passengers each, or 1.4 million passengers each day.[4] Just the waste generated from food service would be a significant target for source reduction. A waste audit conducted at Logan International Airport in Boston showed that approximately one-third of the airport's waste is from the kitchens for in-flight and terminal food service (25 percent is food waste, 40 percent corrugated cardboard, 20 percent mixed paper, and 5 percent glass, aluminum, tin, and other metals).[5]

Among the projects underway at different airlines, Horizon Air and Midway have started serving sandwiches from baskets to eliminate the plastic disposables that are commonly used in airline food services. Japan Airlines washes and reuses plastic chopsticks instead of buying disposable ones. United Airlines has set up a reuse program which donates unopened packaged meals to the Second Harvest Food Bank.[6]

Another idea that could be tried applies to airline shuttle services that provide free newspapers and magazines on racks in their terminals. Signs over the newspaper/magazine racks could ask passengers to leave their publications, once they have read them, on the racks at their destination.

Allied-Signal's Bendix Automotive Systems

Some leading automobile manufacturers have required their suppliers to ship parts in reusable or returnable containers.[7] For example, Allied-Signal's Bendix Automotive Systems plant in Sumter, South Carolina, expects to save $1.1 million between 1990 and 1995 by furnishing collapsible, returnable shipping containers to two of its suppliers.

The Bendix painting plant in Sumter used to receive raw unpainted power booster shells (used in auto and truck brake systems) from its chief suppliers in Ohio and Tennessee in disposable containers. The painted power booster shells were then sent to Bendix customers — automotive manufacturers — in different sets of disposable boxes. Bendix customers include Chrysler, Ford, Jeep, and Hyundai.

To decrease costs and waste, Bendix bought collapsible, returnable containers for two of its suppliers. Now Bendix ships its products to its automotive manufacturer customers in the same containers Bendix suppliers use to ship materials to Bendix. The empty containers can also be collapsed and returned to the suppliers for reuse, thereby completing the circle. These containers replace half of Bendix's expendable corrugated containers. They are painted blue to be distinguishable from other containers.

Bendix reports that the reusables stack higher than the disposable corrugated cases without posing safety hazards, thus using less floor space. As a result, the shops have a neater, cleaner appearance. The reusable containers hold up to 2500 pounds, are strong and lightweight, and collapse to a 3:1 ratio. This allows three times as many to be returned to the supplier in the same truck. Bendix plans to eventually convert all containers to reusables since the beginning of the program has been so successful.[8]

■
Herman Miller, Inc.
Contact
Ted Venti
Address
8500 Byron Rd.,
Zeeland, MI 49464
Phone
616-772-3406
■

Herman Miller, Inc.

Herman Miller, Inc., a major manufacturer of office furniture systems based in Zeeland, Michigan, has pursued a comprehensive packaging reduction program that has saved the company over $1 million each year between 1989 and 1991 and has cut waste in both the packaging coming to Herman Miller and the packaging shipped out by the company.[9]

The waste prevention program aimed at reducing disposable packaging began because customers were demanding that the company's distributors take back the cardboard and plastics in which the furniture was shipped. Herman Miller's waste prevention team discovered that 80 percent of its products were being transported to customers directly from manufacturing facilities and did not need to be completely packaged since the products were shipped in trucks from these facilities. As a result, Herman Miller made changes in its assembly and packaging processes that have eliminated a large portion of the cardboard and plastic that encased orders going in large loads to customers at manufacturers' assembly facilities. These pieces of furniture are instead loaded into trucks and protected with reusable shipping blankets or minimal edge protectors to prevent contact between adjacent pieces. The products are still encased in lightweight plastic bags to keep them clean. Approximately one-third of all office chairs and 95 percent of the wooden desks and files are now shipped in blankets. There were no net cost savings from this particular measure because the reduced costs from using less corrugated were offset by the cost of extra labor for loading the trucks.

Other products, such as the panels used to make office cubicles, are shipped in plastic stretch wrap with protective corners, which uses two-thirds less material by weight than the original packaging. Corrugated expenditures have been reduced by 20 percent overall by utilizing these packaging reduction methods.

An important part of Herman Miller's $1 million cost savings was achieved by source reduction programs directed at materials being shipped to Herman Miller manufacturing facilities. Steel shelves, for instance, are shipped by another company to a Herman Miller manufacturing location. The packaging used to ship these materials generated 16 pounds of waste for every 35 shelves. To reduce this waste, Herman Miller had its suppliers replace the disposable corrugated packaging with reusable protective spacers made out of recycled plastic. This change eliminates 110,000 pounds of waste annually and saves Herman Miller money both through avoided disposal costs and through receiving a price break since the parts supplier has reduced its shipping and packaging costs.

Herman Miller is now looking at how to improve the effectiveness of recycling the remaining packaging materials. The company also recently introduced a program to take back used furniture, refurbish it, and resell it.

Hotels

The United States has more than 44,000 hotels and motels, with a total of some 3 million rooms and more than 1.6 million full- or part-time employees. Lodging industry estimates for waste generation figures are 1 pound per room and 2 pounds per suite on non-checkout days, with these amounts doubling on checkout days.[10]

Many hotels are aggressively pursuing recycling, and some have begun source reduction efforts as well. For example, the Hyatt Hotels Corporation has plans to donate unused food to charities and to farmers who can use it as animal feed, and has proposed replacing disposable paper and plastic cups with a dishwashing system; preliminary figures indicated the $70,000 cost of the system could be recouped in one year.[11]

Additionally, the Boston Park Plaza Hotel and Towers, which is attempting to become an "environmental showcase" through such efforts as recycling, installing water conservation shower heads, and using energy-efficient double-glazed windows, has included some source reduction initiatives. For example, it is banning disposable utensils from its cafeteria and is installing shampoo and lotion dispensers in hotel bathrooms, thereby eliminating the use of 2 million plastic bottles each year.[12]

■
Central Hudson Gas
& Electric Corporation
Contact
Terry Ketcham,
Purchaser
Address
284 South Ave.,
Poughkeepsie, NY
12601
Phone
914-486-5402
■

Central Hudson Gas & Electric Co.

Central Hudson Gas & Electric Co., a utility company in Dutchess County, New York, has been using two-way envelopes for its bills for about 8 years. Approximately 2 million envelopes a year are used to service 250,000 customers.[13] There is a savings of about $0.005 per envelope from not having to purchase a return envelope, which saves $10,000 a year in purchasing costs. It also keeps 2 million unnecessary envelopes out of the waste stream.[14]

Other Business Source Reduction Programs

The American Museum of Natural History in New York City developed a new system that provides members with the option of automatic membership renewal using credit cards. This new system reduces paper waste and allows the museum to save on paper purchases and postage by eliminating the need to mail renewal forms and letters.

Two national supermarket chains, Kroger's and Giant Food, are encouraging source reduction by offering rebates for each paper or cloth bag that is reused. Kroger's gives $0.02 and Giant Food gives $0.03. Rebates are not given for plastic bags since recycling programs are set up for them.[15]

Notes

1. Fibre Box Association, *Fibre Box Handbook,* 1989 Edition, page i.

2. US Environmental Protection Agency, *Characterization of Municipal Solid Waste in the United States: 1990 Update,* June 1990.

3. INFORM (Robert Graff and Bette Fishbein), *Reducing Office Paper Waste,* November 1991, p. 19.

4. Jeanne Trombly, "Airline Recycling Takes Off," *Resource Recycling,* February 1991, p. 62.

5. Amy Martin, "Airport Recycling," *Resource Recycling,* February 1991, p. 64. (Note: 10 percent of the airport's kitchen waste was unaccounted for in the article.)

6. Trombly, *op. cit.,* p. 62.

7. Personal communication, Joe M. Oates, Jr., Xytec, Inc., National Sales Manager, to David Saphire, INFORM, September 9, 1991.

8. "Collapsible Containers to Save $1 Million" *Transportation & Distribution Magazine,* March 1991, p. 62.

9. Information for this section was provided by Ted Venti, Herman Miller, Inc., in personal communications with Caroline Gelb, INFORM, June 1991.

10. Glen Hasek (assistant editor of *Hotel & Motel Management*), "Hotels Keeping Watch on Waste," *Resource Recycling*, January 1991, pp. 56-57.

11. *Ibid.*

12. "Hotel Makeover," *Co-op America Quarterly*, summer 1992, p. 7.

13. Residential customers get six bills a year and industrial customers get twelve bills a year. The average is eight bills per year.

14. Minnesota Office of Waste Management, *Examples of Source (Waste) Reduction by Commercial Businesses*, p. 3.

15. "Reuse Strategy Penalizes Plastic," *Plastics News*, December 9, 1991, p. 43.

Chapter 10 Educating the Public

In 1988, the residential sector generated approximately 108 million tons of municipal solid waste — 60 percent of the US total.[1] Education is the key to encouraging source reduction in this sector.

❖ Educating Consumers

Local governments can publish and distribute a variety of materials that give individual consumers specific recommendations on how to reduce waste. Recommendations can include both purchasing changes and behavioral changes. These educational materials can appear in brochures, pamphlets, booklets, or posters. The more specific the recommendations, the more likely consumers are to take them. For example, a specific recommendation such as "look for longer warranties," or "check and maintain proper air pressure in your tires," is more helpful than generalities such as "buy reusables and repairables," or "maintain your equipment."

Seattle, San Francisco, Berkeley, and New York City are four cities that have developed educational materials that give concrete examples of how individuals can accomplish source reduction in their homes.

Seattle's Use It Again, Seattle! *Directory*

The Seattle Solid Waste Utility published *Use It Again, Seattle!*, a directory of rental, repair, and used goods stores. The guide explains the importance of reducing waste through repairing, renting, and buying used or second-hand goods. It lists repair and rental services for materials ranging from major appliances to exercise equipment, sewing machines, and bicycles. It gives the name, address, and phone number of the different businesses and provides discount coupons from many of the stores as a further incentive to use the services.

■
Seattle Engineering
Department, Solid
Waste Utility
Contact
Carl Woestwin
Address
710 Second Ave.,
Suite 505,
Seattle, WA
98104-1709
Phone
206-684-4684
■

San Francisco's Education Program

San Francisco won an award from the California Department of Conservation for its outstanding education program in 1990. The three major source reduction components of this campaign are public education, environmental shopping, and backyard composting. In the area of public education, the city's Recycling Program worked with Safeway, Inc. (a national supermarket chain) to develop a promotional campaign. Safeway paid for billboards, bus signs, and transit shelter posters that promote the reduction of waste.[2] The Recycling Program also designed an environmental shopping guide that encourages shoppers to buy goods with little or no packaging, reusable or refillable products, and repairable and durable items.[3] Since May 1990, more than 70,000 environmental shopping guides have been distributed at local grocery stores and through the recycling hotline.[4] The third effort is the Recycling Program's brochure on how to compost at home. More than 15,000 brochures have been distributed since 1990.[5]

■
City of Berkeley
Recycling
Address
Martin Luther King Jr.
Civic Center Building,
2180 Milvia St.,
Berkeley, CA 94704
Phone
415-644-8631
■

Berkeley's "Precycling" Campaign

City officials in Berkeley, California developed the term "precycle," which means making purchasing decisions that will have less harmful impacts on the environment. The precycling term is now being used by several other communities, such as Greenwich, Connecticut.

Berkeley's precycling campaign consists of store posters, buttons, and pamphlets with the slogan: "Precycle. Do it right from the start!" The pamphlets give advice to consumers on how to make less wasteful purchasing decisions. The suggestions include:

- Select products carefully and try to buy those that can be reused and have the least packaging.

- Avoid disposables, such as razors, lighters, and plastic plates.

- Buy in bulk to avoid overpackaging and save money; keep perishability in mind to avoid product spoilage.

- Buy durable, long-wearing products, such as premium tires. Use *Consumer Reports* and other consumer publications to research long-lasting products.

■
Department of
Sanitation,
Sanitation Action
Center,
Recycling Office
Address
125 Worth St.,
New York, NY 10013
Phone
212-334-8590
■

New York City's Waste Reduction Handbook

New York City has published a booklet giving consumers specific recommendations on how to be less wasteful shoppers and how to change behavior to reduce waste. The booklet, entitled "New York City's Waste Reduction Handbook: Practical Ways to Prevent Waste and Save the Environment," is available to residents through the Department of Sanitation. Some of its source reduction suggestions are:

- Avoid overpackaged items, such as single serving microwave meals.
- Return hangers to the dry cleaner.
- Use cloth diapers.
- Remove your name from mailing lists.
- Donate items you no longer need, such as books, clothes, and appliances, to organizations that serve the elderly or homeless.
- Reduce reliance on products containing toxic substances. The booklet lists alternative products that can be made at home for declogging drains, cleaning ceramic tiles, polishing furniture, and controlling pests.
- Buy refillables and reusables, such as razors, pens, cameras, and containers for sandwiches (rather than plastic wrap).
- Do not take items you will not use from food stores, such as napkins, disposable utensils, and condiments.
- Bring your own shopping bag to stores.

Reducing Direct Mail

Modern World Design
Address
P.O. Box 249,
New York, NY 10002

Mail Preference
Service, Direct
Marketing Association
Address
11 W. 42nd St.,
P.O. Box 3861,
New York, NY
10163-3861

Governments could also develop and distribute educational materials (fact sheets, brochures, pamphlets, etc.) that target specific items in the waste stream (such as direct mail) or specific businesses. Direct mail is a good example of a specific material to target since it comprises 3 percent of the national waste stream by weight.[6] In 1990, people in the United States received approximately 63 billion pieces of third class mail and threw out about 15 percent of it unopened.[7]

Several brochures already exist that provide people with ways to reduce the mounting problem of direct mail waste. For example, Modern World Design in New York City has designed a "Stamp Out Junk Mail" pamphlet that recommends the following ways to reduce mail waste:

- Write or call companies sending catalogs and direct mail and ask to be removed from unwanted or duplicate mailing lists.
- Ask companies whose catalogs you wish to receive not to sell your name to other companies.
- Contact organizations such as the Direct Marketing Association which provide a service that removes names from direct mail lists. The Direct Marketing Association reports that more than 19 million people have asked to be removed from direct mail advertising lists.[8]
- Write to the Postal Rate Commission to request that higher rates be established for bulk mail.

❖ Educating Students

Programs for school-age children can interest them in living less wastefully. They can often, then, become enthusiastic advocates, encouraging others to also develop less wasteful habits.

Nevertheless, as noted in the Chapter 4 section on accomplishing source reduction in schools, most garbage management efforts in schools focus on recycling. This is as true for school curricula as it is for school operators. The three education programs discussed below — in Minnesota, Wisconsin, and Westchester County, New York — provide examples of the few activities directly aimed at teaching source reduction in schools.

There are many more possibilities for incorporating source reduction activities directly into school curricula than are included in these three programs. Some additional ideas are also listed here.

Minnesota

■
Minnesota Office of
Waste Management,
Waste Education
Contact
Ruth Marston
Address
1350 Energy Lane,
St. Paul, MN 55108
Phone
612-649-5738,
800-652-9747
(toll free in MN)

Clearinghouse:
612-649-5482;
800-877-6300
(toll free in MN)
■

The Minnesota Office of Waste Management (OWM) is developing a curriculum for kindergarten through 12th grade. The curriculum will cover solid and hazardous waste issues and will be developed as an "Outcome Based Education" (OBE) system. OBE is a form of education that teaches students why learning specific information is necessary and how it can be applied to real life activities. The consultant designing the curriculum is required to follow the waste management hierarchy of "reduce and reuse first."

A pilot curriculum for the third grade is already being used throughout the state, after field testing in 14 schools. The curriculum, "Community Cats," was used in the 1989-1990 school year. Despite the fact that the guidelines for the full-scale curriculum place source reduction first, the pilot focused on recycling. It did, however, include a few source reduction ideas. It suggested that schools have swap boxes in the classroom for students to exchange unwanted items from home and that children donate old belongings to parents' garage sales or donation organizations, such as Goodwill.

The Office of Waste Management hopes to receive funding to conduct teacher training workshops, give curricula to all teachers who request it, and make the material available on computer disks. The program was scheduled to be available for all grades in schools statewide by fall 1992.

Lists of other waste education curricula can be obtained through Minnesota's waste education clearinghouse which stores a variety of information on solid and hazardous waste issues.

Wisconsin

Wisconsin's waste education curriculum consists of five guides: *Recycling Study Guide, K-3 Supplement to the Recycling Study Guide; Nature's Recyclers:*

■
Recycling Education
Coordinator,
Wisconsin Depart-
ment of Natural
Resources
Address
P.O. Box 7921,
Madison, WI 53707
Phone
608-266-2742
■

*Activity Guide; The Fourth "R": An Action Booklet for Recycling in the Class-
room and School; and Nature's Recyclers Coloring Book.* Although the titles
imply concentration on recycling, the books include several strong source reduction activities. Among these activities are:

- Discussions of nonrenewable and renewable natural resources, waste
  management options, and costs.

- Student collection of daily trash to help them visualize the amount generated and develop opportunities for elimination, reduction, and reuse.

- Student activities to reduce the amount of paper, food, packaging, and
  other materials used during the school day, such as using paper scraps in
  class and bringing food in reusable containers.

- Backyard composting/decomposition projects.

- Packaging discussions including developing labeling programs, and
  marketing and advertising ideas, and designing packaging.

- Daily source reduction and recycling classroom practices (collecting unwanted pens and notebooks at the end of the year for distribution the
  following year; keeping a swap box for toys, books, and games in a central location; and increasing use of overhead transparencies and bulletin
  boards instead of handouts).

Wisconsin's education program is designed so that some activities are for one-
time class use and others are to be included in the class syllabus, thus being
integrated into everyday learning activities. The activities can be incorporated in
various classes such as science, math, English, and environmental studies.

Westchester County, New York

■
Board of Cooperative
Educational Services,
Instructional Services
Division
Address
2 Westchester Plaza,
Elmsford, NY 10523
Phone
914-345-8500
■

Westchester County, New York, is promoting a waste reduction curriculum that
was compiled by the Southern Westchester Board of Cooperative Educational
Services (BOCES). The curriculum combines various activities and worksheets
from other state programs, as well as ideas from organizations such as the
Audubon Society and the Sierra Club. This curriculum shows that waste education
programs do not have to be developed on the state level and do not have to cost a
lot of money because they do not have to be designed from scratch.

Westchester's program covers source reduction as well as recycling issues and
is split into four sections: reduction, reuse, natural recycling (composting and
biodegrading issues), and "mechanical" recycling (the actual technical process).
The curriculum provides background and activities (fun and practical). The "fun"
activities include developing recycling slogans and making musical instruments
out of materials that would normally be discarded, such as milk cartons and plastic
wrap. The practical activities involve actual source reduction, recycling, and
composting, such as mini-composting experiments that show the benefits of

backyard composting, and garbage surveys to identify disposal habits and opportunities for reduction.

The curriculum was presented at BOCES' first environmental forum for educators in celebration of Earth Day 1991. Fifty-five teachers from schools throughout the county attended and were taught how to introduce the curriculum to other teachers and to students. Education in schools was further promoted at the 1992 Earth Day events.

Other Curriculum Ideas

Other ideas for incorporating source reduction into school activities include:

- Establishing an extracurricular club that focuses on developing and implementing source reduction practices within the school.

- Setting up suggestion boxes so that teachers and students can relay additional source reduction ideas and identify any problems with the program.

- Teaching the difference between recycling and source reduction.

- Developing class projects to identify less wasteful products and strategies for encouraging consumers to use them.

- Developing worksheets and videos.

- Sponsoring source reduction contests for posters and essays that demonstrate ways to prevent waste generation.

- Using items in arts and crafts projects that would otherwise be discarded.

Notes

1. Franklin Associates, Ltd., estimate. Personal communication: Nicholas Artz, Franklin Associates, to Caroline Gelb, INFORM, March 1992.

2. Amy Perlmutter and Maria Hon, "San Francisco's Comprehensive Recycling Program," *Public Works Magazine,* July 1991.

3. San Francisco Recycling Program, "Environmental Shopping Guide," 1990.

4. Perlmutter and Hon, *op. cit.*

5. *Ibid.*

6. Jill Smolowe, "Read This!!!", *Time,* November 26, 1990, p. 64.

7. Michael W. Miller, "'Greens' Add to Junk Mail Mountain," *The Wall Street Journal,* May 13, 1991.

8. "Programs in Action," *Resource Recycling,* August 1991, p. 20.

■ Chapter 11 Economic Incentives and Disincentives

Local governments can provide economic incentives or disincentives that encourage manufacturers and consumers to produce less waste. Such programs include variable waste disposal fees, advance disposal fees, taxes, tax credits, deposit/refund systems, and financial bonuses.

❖ Variable Waste Disposal Fees

Some localities across the country are adopting variable waste disposal fees — systems that charge residents directly based on the amount of waste set out for disposal. Such systems are now in place in more than 200 communities in 19 states.[1] This is in contrast to the traditional method of funding garbage management through general tax revenues. Variable waste disposal fees create an incentive to reduce and recycle since the more garbage a household generates, the more it must pay for disposal. Generally, a fee is charged for garbage, but recyclables are picked up free. While these variable rate systems are now being used to encourage reductions in waste, some cities have been using them for years as their standard way of paying for garbage disposal. San Francisco has been using such charges for garbage service since 1932, and Olympia, Washington, since 1954.

Charging users directly for services provided by governments to create an incentive to conserve resources is not a new idea. Most cities in the United States, for example, use a water metering system to charge residents for the amount of water they use; the need for water conservation is one factor motivating the installation of water meters. Studies by New York City, which is starting such a system for single-family homes, show that direct charges for water through a metering system can reduce water consumption by 10 to 30 percent.[2]

Variable waste disposal fee systems are generally based on volume. Currently, the two main systems are a pre-paid tag or sticker system and a per can volume-based system. With the tag or sticker programs, residents purchase bags, tags, or stickers at local stores or other designated spots and place them on bags that are to be collected. Garbage is only picked up if the official bag or tag is used. Under the per can system, residents are charged for the number of cans (the charge may differ depending on the can size) set out for collection.

Some communities (including Seattle; Durham, North Carolina; and Farmington, Minnesota) are beginning to explore weight-based systems because volume-based systems have led to compacting and are based on the size of the container rather than the actual garbage put out. For example, the charge for a 30-gallon container is the same if it is half full or completely full.

The weight-based systems would equip trucks with scales and bar coding equipment that could weigh garbage at curbside and record the amount discarded by each household. While the cost of this equipment is now $5000 to $10,000 per truck, this is expected to decrease if such systems become more widespread.[3] Two advantages of such weight-based systems are, first, that they provide a greater incentive to reduce waste since the customer is charged for the actual weight of waste discarded (not the size of the bag or can, which may or may not be filled) and, second, that they can provide useful data for measuring source reduction.

Implementation of variable rates involves several steps:

- Determining whether state law empowers the local agency to bill for solid waste on the basis of amount;

- Developing the rate system;

- Establishing ordinances to require the charge for solid waste management;

- Prohibiting illegal dumping, establishing enforcement mechanisms, and educating residents to comply;

- Adapting trucks for a weight-based system;

- Adapting systems for multi-tenant buildings that limit illegal disposal;

- Restructuring for low-income households since mandatory fees place inequitable burdens on low-income customers.[4]

Although many environmentalists and government officials hail variable waste disposal rates as promoting source reduction as well as recycling, little data are available to show how effective this system is at accomplishing source reduction. In fact, as long as recyclables are collected at no charge, these programs do not encourage source reduction more than recycling. However, these systems can be adapted to place more focus on reduction than recycling. Charging for the amount of recyclables, but at a lesser rate than for nonrecyclable garbage, could promote source reduction. Education about source reduction could help residents determine ways to reduce recyclables and garbage as well as save money. Variable rates certainly increase residents' awareness of the garbage they produce; this is the first step in accomplishing source reduction.

Two examples of well developed variable waste disposal rate programs are in Tompkins County, New York, and Seattle, Washington.

Tompkins County, New York

■
Tompkins County
Solid Waste Division
Contact
Barbara Eckstrom,
Solid Waste Manager
Address
Bostwick Road,
Ithaca, NY 14850
Phone
607-273-6632
■

Tompkins County, New York, which includes the city of Ithaca, operates the Tompkins County Trashtag Program. Residents buy tags from licensed haulers to put on trash bags; the tags specify the weight allowance of each bag set out for disposal. The tagged bags are only for garbage. Recyclables are put in separate, untagged containers, and are not charged for separately.

A reduction has been observed in the amount of waste set out for disposal since the program began in February 1990, but actual amounts of reductions have not been measured. There is no mechanism to determine which reductions are due to recycling or source reduction, and which to compaction. The haulers have the right to refuse a pickup if the bag weighs more than the amount allowed by the particular tag. However, bags are not weighed; haulers estimate their weight.

To evaluate the effectiveness of its tag and recycling programs, the Tompkins County Solid Waste Management Division, in conjunction with the Waste Management Institute at Cornell University, conducted a county-wide survey. The survey consisted of 76 questions pertaining to how much waste residents believe they generate (weight and number of bags), whether materials are recycled through curbside or drop off, what types and what amounts of certain materials are recycled, composting habits, consumer awareness of source reduction, and demographics. The results of the survey show that 47 percent of survey respondents believed they were generating less waste than before the Trashtag Program.

The survey also revealed what people said about their garbage practices and opinions; some of these results are listed below.

Disposal:

- 20.4 percent burn some of their garbage
- 58.7 percent compact some of their trash
- 51.1 percent believe there is more illegal dumping and burning as a result of the Trashtag Program

Composting:

- 7.6 percent more residents compost than did so before the Trashtag Program
- 15.5 percent of those who were doing some composting before the Trashtag Program are now composting more

Purchasing:

- 44.8 percent occasionally choose products they believe to have less packaging[5]

■
Seattle Engineering
Department, Solid
Waste Utility
Contact
Carl Woestwin
Address
710 Second Ave.,
Suite 505,
Seattle, WA
98104-1709
Phone
206-684-4684
■

Seattle, Washington

Seattle is the only city INFORM found that has reported results from its variable waste disposal system. Ratepayers select subscription levels based on the number of cans of garbage they generate each week. The city offers subscription levels in standard 30-gallon increments (1 can, 30 gallons; 2 cans, 60 gallons; etc.), as well as a "mini-can" or "zero can" that holds 19 gallons, which is called the "super-recycler" option. The rates the customers pay increase significantly with additional numbers of cans. As of July 1992, the monthly single family rate was $11.50 for placing one mini-can out each week, $14.98 for one full-size can per week, and $14.98 for each additional can.[6] Separate rates apply to low-income rate payers: $4.60 a month for one mini-can per week, and $5.99 for one full-size can.

The program has been so successful that customers are asking for cans smaller than the mini-can for weekly waste. Since 1981, the average subscription for a residential ratepayer has fallen from 3.5 to 1.4 cans. However, cans set out for recycling are not included in this statistic, and it does not account for compaction. Thus, it is unclear whether the results reflect just increased recycling and/or compaction, or also actual waste reduction at the source. According to the Seattle Solid Waste Utility, there is certainly better separation of recyclables, and people are more aware of what they are throwing away, thus creating a climate in which source reduction could be increasingly encouraged.[7]

As noted at the beginning of this section, Seattle is beginning to explore the option of a weight-based garbage system that would allow each customer to be charged on the basis of the weight of the garbage put out for collection, thus neutralizing the effects of different compaction rates and partially empty cans. This system would also be useful for Seattle because its landfill fees are weight-based.

❖ Taxes/Fees

Taxes can encourage consumers to make purchasing decisions that have less harmful impacts on the environment.

Taxes or fees designed to incorporate the cost of waste disposal into the price of the product or package are called advance disposal fees (ADFs). A report by the Arthur D. Little consulting firm cites 28 state legislative bills proposing ADFs that were being considered in 1991.[8] These taxes and fees may be imposed at any point in the product manufacture, packaging, or distribution chain: on the manufacturers, wholesalers, retailers, or consumers. The taxes or fees can be based on weight, volume, or price. To date, for administrative simplicity, most have been levied per item. Existing and proposed ADFs are too low to have a significant impact on consumer behavior and are primarily a source of revenue.[9]

Table 11-1 lists states that have passed or considered product and packaging tax legislation. Two states have placed taxes on lead-acid batteries, 18 states have

Table 11-1 *Fees Imposed on Products and Packages in the United States (as of mid-1992)*

	Lead-Acid Batteries (tax per unit)	Diapers (tax per unit)	Tires (tax per unit)	Other
Arizona			2% of tire cost*	
California			$0.25	
Florida	$1.50		$1	$0.01 per container on glass, metal, and plastic†
Idaho			$1	
Illinois		$0.01	$0.50	
Kansas			$0.50	
Kentucky			$1	
Louisiana			$2	
Maine	$1		$1	$5 major appliances‡
Missouri			$0.50	
Nebraska			$1	
North Carolina		1%		
Oklahoma			$1	
Oregon			$0.50	
Rhode Island				$0.05 per quart of motor oil, $0.10 per gallon of antifreeze and organic solvents
Utah			$1 to $2**	
Texas			$2	
Virginia			$0.50††	
Wisconsin			$2	

*   Tax cannot exceed $2 per tire.
†   If a 50 percent recycling rate is not achieved for these containers by 1992, the fee will increase to 2 cents per container.
‡   This refers to new major furniture, appliances, bathtubs, and mattresses.
** Exemption for recycled content and bicycle tires.
†† Exemption for bicycle tires.

placed taxes on tires (with action in one state pending), one state has placed a tax on disposable diapers, and three states have placed taxes on other items such as major appliances, packaging materials, and motor oil. In all, 19 states have placed fees on at least one material.

Taxes on products or packaging that encourage recycling but not source reduction could be modified to encourage source reduction first. For example, Florida, which imposed a disposal fee on newspaper publishers that gives a credit for newsprint returned for recycling, could give an additional credit to newspaper publishers for reducing the amount of newsprint returned for disposal and recycling.

❖ Deposit/Refund Systems

Deposit/refund laws place fees on products when they are purchased and give the fee back to the consumer upon return of the container or product. Governments currently aim deposit/refund systems at recycling, but few use them to promote source reduction. They could do so by encouraging local manufacturers to develop deposit/refund systems for reusable items, supporting state and national efforts to develop deposit/refund systems on a larger scale, and educating consumers to participate in these systems.

Deposit systems for beverage containers (bottles and cans), commonly called bottle bills, are the most common form of deposit/refund systems in place in the United States. Currently, nine states have beverage container deposit legislation: Connecticut, Delaware (not including cans), Iowa, Maine, Massachusetts, Michigan, New York, Oregon, and Vermont. California has a 5 cent redemption law, but deposits are not required.

Bottle bills have return rates of between 72 and 98 percent. Most of the bottles collected in this way are recycled; few are refilled.[10] Michigan and Oregon are the only states that attempt to promote refillable over one-way containers, thus promoting source reduction over recycling. In Michigan, bottles that can be refilled carry a 5 cent deposit, while one-way bottles and cans have a 10 cent deposit. Oregon requires a 2 cent deposit on refillables and a 5 cent deposit on one-way beverage containers. Overall, the market share of refillable beer bottles ranges from 7 to 24 percent in the nine states with bottle bills;[11] in Michigan, the rate was 14 percent in 1990 (down from 21 percent in 1982), and in Oregon, the rate was 7 percent in 1990 (down from 27 percent in 1982). The national average is approximately 6 percent for both beer and soft drink containers, down from 46 percent for beer and 89 percent for soft drinks in the early 1960s.

Eight of the ten states with the highest market share of refillables for beer are, in fact, bottle bill states, according to Beer Institute figures. This might be explained by the fact that bottle bill states have a return infrastructure in place that aids in returning empty bottles to the brewery for refilling. Two sister breweries owned by G. Heileman provide an illustration of differences in bottle bill and non-bottle bill states. Blitz-Weinhard in Oregon, a bottle bill state, gets back approximately 75 percent of its empties. Rainier, in the adjacent non-bottle bill state of Washington, only gets back 25 to 30 percent.[12]

Bottle bills alone may indirectly promote refilling because return systems are set up to meet legislative requirements. However, to have a major effect, bottle bills may have to be specifically aimed at promoting refills.

❖ Tax Credits

Governments can give tax credits to businesses and institutions that take steps to reduce waste at the source in the commercial sector, although INFORM's research for this project did not identify any governments that have done so. For example, credits can be given to manufacturers that:

- Reduce packaging
- Change to reuse/refill of beverage containers or other consumer products
- Increase a product's durability
- Eliminate or reduce toxic constituents in a product or in its packaging

Credits can also be given to businesses that buy waste-reducing equipment such as double-sided copy machines, reusable tableware in cafeterias, and plain paper fax machines.

❖ Financial Bonuses

At the disposal level, financial bonuses can be given that encourage source reduction. For example, Westchester County, New York, compares the amount of waste each locality sends to disposal (incineration) to the amount sent in the previous year and pays the locality $17 for each ton reduced. This program, however, equally rewards recycling and source reduction.

A tonnage grant program in New Jersey, by contrast, promotes recycling at the expense of source reduction. The state pays localities for each ton of materials collected for recycling with grants ranging from $0.98 to $1.82 per ton, depending on who does the recycling. This system could lead local recycling coordinators to oppose source reduction efforts which would, if successful, reduce the grants by reducing the amounts of materials generated and recycled.

Notes

1. Lisa Skumatz, "Garbage by the Pound," presentation at Second US Conference on Municipal Solid Waste Management, June 3, 1992.

2. Personal communication, Hilary Dobies, New York City Department of Environment Protection, to Caroline Gelb, INFORM, October 31, 1991.

3. Lisa Skumatz, "Garbage by the Pound," *op. cit.*

4. Lisa A. Skumatz and Cabell Breckinridge, *Variable Rates in Solid Waste: Handbook for Solid Waste Officials: Volume I - Executive Summary, Future Impressions,* Seattle, June 1990, pp. 10-25.

5. Sarah Stone, *A Final Report: Tompkins County Trashtag and Recycling Study,* prepared for Cornell Waste Management Institute and Tompkins County Division of Solid Waste, 1991, pp. 1, 12-15.

6. Personal communication, Carl Woestwin, Seattle Solid Waste Utility, to Caroline Gelb, INFORM, July 1992.

7. "On The Road to Recovery," *Seattle's Integrated Solid Waste Management Plan,* August 1989, p. 28.

8. Arthur D. Little, Inc., *A Report on Advance Disposal Fees,* prepared for Environmental Education Associates, April 1992, pp. 1-12.

9. Studies by Tellus and Arthur D. Little estimate that advance disposal fees that would capture the cost of disposal would average roughly 1 percent of the price and that consumer elasticity of demand is less than 0.5; therefore, such fees would reduce purchases by less than 0.5 percent.

10. United States General Accounting Office, Report to Congressional Requesters, "Solid Waste: Trade-offs Involved in Beverage Container Deposit Legislation," November 1990, p. 34.

11. Beer Institute, *1990 and 1982 Estimate Draught and Container Share by State,* Washington, DC.

12. Rainier Brewing Company Press Release, "Refill or Landfill," April 15, 1991

▓ Chapter 12 Regulatory Measures

Regulations can be adopted on the local, state, or federal levels that mandate or encourage reduced waste. They can take a variety of forms: requirements that businesses conduct waste audits and develop source reduction plans; labeling schemes that give consumers information to facilitate purchasing decisions that have less harmful effects on the environment; bans on the sale or disposal of specific materials or products; and legislation requiring manufacturers to reduce packaging or decrease the toxic constituents in products.

❖ Required Source Reduction Plans

Local governments can require businesses and institutions to conduct waste audits and materials assessments and develop source reduction plans. Conducting an audit and writing a plan reveal to businesses opportunities not only to reduce waste but often also to save money. While many states have required their toxic waste-generating industries to plan for and report reductions in plant production wastes, only Rhode Island and New Jersey have established such requirements in the area of commercial solid waste.

Rhode Island

Rhode Island required every business with more than 50 employees to submit a recycling and waste reduction plan by the end of 1991, and to update it annually. Of the approximately 28,000 businesses in Rhode Island, only 1200 have more than 50 employees. The businesses were required to evaluate reduction opportunities before stating their recycling plans. However, the initial plans submitted focused mainly on recycling. According to the Rhode Island Department of Environmental Management (DEM), once recycling programs are in place, future plans will probably focus more on source reduction.[1] The Department of Environmental Management plans to evaluate the businesses' annual reports to quantify

■
Rhode Island Department of Environmental Management, Division of Environmental Coordination
Contact
Susan Cabeceiras, Commercial Recycling Coordinator
Address
83 Park St., Providence, RI 02903
Phone
401-277-3434
■

the information from the plans, estimate cost savings and expenditures, and note trends and dramatic individual records.

Despite the original focus of both the state and its businesses on recycling, the Rhode Island requirement has led some businesses to adopt source reduction strategies. One jewelry business switched from using corrugated cardboard (a material that must be recycled under state law) to reusable containers within its facility.

New Jersey

■
New Jersey Department of Environmental Protection and Energy, Division of Solid Waste Management
Contact
Athena Sarafides, Waste Reduction Specialist
Address
840 Bear Tavern Rd., CN 414, Trenton, NJ 08625-0414
Phone
609-530-8208
■

Recommendations released by the New Jersey Governor's Emergency Solid Waste Assessment Task Force in August 1990 urged counties to require industries to conduct waste audits and develop waste reduction plans. However, no enforcement mechanism was included. The Task Force recommended the following due dates for the plans: 1992 for industries with more than 500 employees, 1993 for those with 250 to 499 employees, and 1994 for those with 100 to 249 employees (most New Jersey businesses fall into this category). As of the summer of 1992, three of the twenty-one counties had indicated that they will conduct waste audits in the private sector.

❖ Labeling

As more and more consumers voice concerns about the environmental impacts of products, labeling programs are being designed to help them determine the environmental effects of products and packages. In some cases, products are labeled directly; in others, labels are placed on store shelves.

Many factors are involved in the environmental impacts of products, making it difficult to quantify what makes one product "better" or more "environmentally friendly" than another, overall. As discussed in Chapter 2, efforts are underway to develop complete product lifecycle analyses aimed at more systematic product assessment.

Since so many products are marketed nationally, labeling requirements and definitions for different states and localities are not practical for either manufacturers or consumers. Manufacturing costs for meeting different state labeling requirements would be high, and a variety of labels would be confusing to consumers. Thus, labeling standards would be most appropriately regulated at the federal level. However, no federal standards have emerged in the United States.

Labeling for source reduction is feasible and is being implemented abroad. In Germany, companies apply for the environmental approval label, the "Blue Angel," for their products and packaging. Germany set criteria for 50 product and packaging categories. If the product or packaging meets the criteria, the Blue Angel is awarded. Of the 50 categories, 19 are in the area of source reduction, such as reusable food containers, reusable transport packaging, and water-based paints.

The same process is used in Canada's environmental labeling program. As of 1992, Canada had 16 product and packaging categories for which criteria have been set. Six of these are directed at source reduction, including categories such as cloth diapers and reusable shopping bags.

In the United States, most environmental labeling efforts have been focused on recycled content and recyclability of products and packages. The Environmental Protection Agency and the Food and Drug Administration are developing nationwide environmental labeling standards, and private organizations such as Green Cross and Green Seal are conducting lifecycle analyses of individual products to determine whether or not to award them a seal of environmental approval.

One program, in Iowa, involves labeling at the local level; another, involving toxic constituents in products, is described in Chapter 13.

Grinnell, Iowa

■
Grinnell 2000
Foundation
Contact
Donna Sullivan or
Greg Prestemon
Address
P.O. Box 771,
Grinnell, IA 50112
Phone
515-236-6311
■

The Grinnell 2000 Foundation in Grinnell, Iowa, is conducting a labeling program that places stickers on products to show whether their packaging meets waste reduction criteria. The program, funded by the state Department of Natural Resources, is running in a chain of grocery stores — four in the city and two in small neighboring towns. Product packaging is evaluated for source reduction, recyclability, and recycled content. The product or package receives a "check" for source reduction if it uses what the Grinnell Foundation has determined is the "least" packaging — that is, if the product's packaging will result in less waste than that of other commercially available products of the same category. (The Grinnell Foundation has not clearly defined the criteria it uses for determining if the least amount of packaging is used.) Other grocery chains have expressed interest in adopting this labeling scheme.

The program also includes consumer education. Posters and brochures promoting waste reduction are available at stores, and a handbook was distributed to all residents that provides practical guidelines for source reduction and recycling. As of mid-1992, the Grinnell Foundation had not evaluated the results of the program.

❖ Bans

Thirty-three states and the District of Columbia have adopted bans on the disposal of specific materials (see Table 12-1). Such bans can encourage manufacturers to develop alternatives that create less waste because consumers may stop buying items they cannot dispose of in their trash. Some states and several cities have also instituted retail bans: banning the sale or use of specific products and packaging. Bans can be reinforced with consumer education to encourage source reduction.

Disposal Bans

Currently, disposal bans are generally designed to increase recycling rates of certain materials and to keep materials out of disposal systems because of their potential harm to the environment. Lead-acid batteries, for example, can ooze lead and oxides of sulfur and contaminate water supplies if landfilled, and can generate emissions of lead if burned. Banning these batteries from disposal increases the incentive to return them to retailers for recycling.

INFORM's research on the 50 states and the District of Columbia found that 34 states and the District of Columbia have banned materials from disposal: 16 have banned yard waste, 30 have banned lead-acid batteries, 11 states and the District of Columbia have banned used oil, 18 have banned tires, and 7 have banned major appliances (such as refrigerators and stoves). **Table 12-1** lists the specific items each state bans.

To promote source reduction when banning certain materials from disposal, governments can publicize information about alternatives. For example, a ban on disposal of yard waste could be coupled with an educational campaign to teach people how to compost waste in their backyards.

Eight of these states also ban other materials from disposal.

- Connecticut bans nickel-cadmium batteries.

- Florida bans construction and demolition debris.

- Iowa bans nondegradable grocery bags and bottle bill containers.

- Massachusetts bans glass and plastic containers and metal from disposal.

- Minnesota bans dry cell batteries that contain heavy metals, such as mercuric oxide or silver oxide electrodes and nickel-cadmium.

- Oregon will not let recyclables that have already been separated from other garbage enter its disposal facilities.

- Rhode Island has passed legislation to keep recyclables out of disposal facilities.

- Wisconsin bans aluminum containers, corrugated paper or boxes, glass containers, newspapers, office paper, plastic packaging, steel containers, and magazines.

Retail Bans

Retail bans have been instituted in several localities around the United States. For example, Portland (Oregon), North Carolina, Suffolk County (New York), and Newark (New Jersey) all have banned plastic bags and containers from food stores and restaurants. These measures promote reductions in the toxic constituents related to plastics packaging, but not necessarily reductions in the amount of

Table 12-1 Disposal Bans in the United States (as of mid-1992)

State	Yard Waste	Lead-Acid Batteries	Used Oil	Tires	Major Appliances	Other
Alabama						
Alaska						
Arizona						
Arkansas	■	■		■		
California		■				
Colorado						
Connecticut		■	■			Nickel-cadmium batteries
District of Columbia			■			
Delaware						
Florida	■	■	■	■	■	Construction and demolition debris
Georgia		■				
Hawaii		■				
Idaho				■		
Illinois	■	■		■		
Indiana		■		■		
Iowa	■	■	■	■		Bottle bill containers, household batteries
Kansas		■		■		
Kentucky		■				
Louisiana		■				
Maine		■				
Maryland						
Massachusetts	■	■	■	■	■	Glass and plastic containers, metal
Michigan	■	■	■			

continued

Table 12-1 *Disposal Bans in the United States (as of mid-1992) (continued)*

State	Yard Waste	Lead-Acid Batteries	Used Oil	Tires	Major Appliances	Other
Minnesota	■	■	■	■	■	Dry cell batteries with heavy metals
Mississippi						
Missouri	■	■	■	■	■	
Montana						
Nebraska						
Nevada						
New Hampshire		■				
New Jersey	■*					
New Mexico						
New York		■				
North Carolina	■	■	■	■	■	
North Dakota						
Ohio	■	■		■		
Oklahoma				■		
Oregon		■		■		Separated recyclables
Pennsylvania	■†	■				
Rhode Island						Recyclables
South Carolina						
South Dakota						
Tennessee		■	■	■		
Texas	■	■				
Utah						
Vermont		■	■	■	■	

continued

State	Yard Waste	Lead-Acid Batteries	Used Oil	Tires	Major Appliances	Other
Virginia		■				
Washington		■				
West Virginia						
Wisconsin	■	■	■	■	■	Containers (glass, aluminum, steel), corrugated newspapers, office paper, magazines, plastic packaging
Wyoming	■	■				

* Bans only leaves during fall months.
† Bans only leaves.
Note: Some of these bans have not yet gone into effect.

garbage because other disposable products, such as paper bags or aluminum food containers, may be substituted. Berkeley, California, has banned the use of foam products in government buildings.

In order to achieve reductions in amounts of waste, governments could couple retail bans on disposable products and packaging with consumer education and economic incentives to utilize reusable products instead. One idea: governments could supplement a ban on polystyrene with consumer education about reusable mugs and could encourage retailers to offer incentives to customers who bring in their own mugs for take-out coffee or reuse other containers for take-out food. Stores would benefit if customers used their own reusable containers, since they would save the costs of purchasing and storing disposables; government could encourage them to pass along some of the savings to the customers.

The Environmentally Acceptable Packaging Ordinance in Minneapolis and St. Paul, Minnesota, includes a source reduction provision. In addition to banning the sale of nonrecyclable packaging, it requires all restaurants, grocery stores, and other food service operators to sell and use packaged items only if their packaging is either reusable or recyclable.

❖ Packaging

At one-third of the waste stream, packaging is a prime target for source reduction. United States residents generate an average of 463 pounds of packaging waste per person per year — far more than residents of many other industrialized countries, as illustrated in **Table 12-2.** Significant reductions in waste generation and costs could be achieved by eliminating unnecessary packaging, reusing/refilling, and designing more efficient packages.

Yet the whole issue of how to reduce packaging waste is fraught with controversy: there is little consensus about what packaging is "essential" and what is "excessive." Packaging, in fact, serves many purposes. It is used for protecting product safety, containing products, transporting products, and marketing products. Further, different levels of packaging are required for different types of products. Pharmaceuticals, for example, may require more packaging for safety purposes than would be necessary for other products.

Marketing-related packaging is the primary area of controversy. Manufacturers often consider reducing such packaging a threat to the marketability of their products. Yet, if there were established guidelines or regulations about such packaging, all manufacturers would be "playing by the same rules" as their competitors.

Broad policies to reduce packaging waste by making the companies that produce, use, and sell packaging responsible for managing the waste can most effectively be implemented by the federal government since this requires dealing primarily with national companies and a national distribution system. To date, the federal government has taken little action to reduce the amount of packaging waste. However, as of mid-1992, legislation was pending in Congress, as part of the reauthorization of the Resource Conservation and Recovery Act (RCRA), to reduce the use of four heavy metals — lead, cadmium, mercury, and hexavalent chromium — in packaging, a proposal modeled after one originally advanced in 1990 by the Coalition of Northeastern Governors (CONEG). The CONEG proposal has been adapted to date by 11 states — Connecticut, Iowa, Maine, Minnesota, New Hampshire, New Jersey, New York, Rhode Island, Vermont, Washington, and Wisconsin. (See Chapter 13 for further discussion of reducing toxic chemicals in products at the source.)

Table 12-2 *Packaging Waste in OECD Countries, late 1980s*

Country	Amount per Capita (pounds)
Belgium	595
Canada	485
United States	463
France	399
Japan	359*
Netherlands	344
United Kingdom	295
Austria	291
Finland	289
Germany	276
Italy	242
Australia	220†

* Represents consumption of packaging, not waste; therefore, waste data would be substantially lower.

† Represents residential waste only.

Source: Congressional Research Service Report for Congress (James E. McCarthy), Recycling and Reducing Waste: How the United States Compares to Other Countries, November 8, 1991, p. 8.

Current State and Local Packaging Initiatives

Packaging legislation is currently being introduced in state legislatures across the country. One bill, developed by the Massachusetts Public Interest Research Group (MassPIRG) and introduced in Massachusetts, has been a model for state bills proposed in Florida, New Hampshire, New Mexico, New Jersey, New York, North Carolina, and Pennsylvania. The Massachusetts bill requires that, by 1996, all packaging used in Massachusetts be either reusable five times, made of 50 percent recycled material, or made of materials being recycled at a rate of 35 percent in the state. Such bills are often referred to as "rates and dates" legislation. An Oregon bill, passed in July 1991, limits coverage to glass and plastic packaging.

No states or localities have taken action to reduce the amount of packaging waste by banning or taxing packaging, mandating refillables, or eliminating distribution of free bags at retail stores. While "rates and dates" laws will reduce the amount of packaging waste requiring disposal, the proposed bills do not make source reduction the top priority. By mixing recycling and source reduction goals, they allow recycling to be substituted for source reduction.

Packaging legislation involving bans has also focused on increased recyclability or on specific problem materials rather than on source reduction. For example, 25 states have banned nondegradable plastic beverage container rings and eight states have banned plastics blown with chlorofluorocarbons (CFCs). There are a variety of bans on plastic bags or packages that are not degradable or recyclable. Maine bans aseptic packages; however, the ban contains no provisions mandating that these be replaced by packages that create less waste (although they will, presumably, be more recyclable).

The Coalition of Northeast Governors

The Coalition of Northeast Governors (CONEG) formed a Source Reduction Council in 1989 to advise the governors from the nine states in that region (Connecticut, Maine, Massachusetts, New Hampshire, New Jersey, New York, Pennsylvania, Rhode Island, and Vermont) on strategies to reduce packaging waste. This council, which developed the model legislation to reduce heavy metals in packaging described above, was originally composed of representatives from state government, private industry, and nonprofit environmental organizations (including INFORM). The environmentalists ultimately withdrew from the council.

Members of the Source Reduction Council were unable to agree on model legislation to reduce the amount of packaging waste. Instead, CONEG adopted "Preferred Packaging Guidelines" that suggest steps industry can take on a voluntary basis to reduce packaging waste. A "CONEG challenge" was issued, urging companies to implement the "Preferred Packaging Guidelines." As of November 1991, 30 major companies had accepted the challenge.

While the CONEG guidelines advocate minimal packaging and reusing/refilling, they do not set numerical goals to be achieved by specified dates, nor do they establish separate source reduction and recycling goals. For example, under the CONEG guidelines, companies may opt for packaging with redundant layers, provided the packaging is made of recycled content and/or is recyclable. A source reduction strategy would aim to eliminate the extra layers.

Before withdrawing from CONEG, the environmental organizations developed a packaging agenda that set a goal of reducing packaging waste by at least 40 percent from current levels within 5 years, independent of recycling. A three-pronged approach was suggested that would (1) eliminate excessive packaging, (2) promote returnable/refillable packaging, and (3) promote the use of bulk retail packaging. Specific strategies included drafting model legislation to:

- Ban "egregious" packaging; the definition of this was left up to CONEG and, as noted earlier in this section, such definitions are one of the most problematic aspects of dealing with packaging.

- Implement a packaging efficiency standard based on a product-to-package ratio.

- Set aside a percentage of retail shelf space for refillable/returnable packaging.

- Require distributors of liquid products to deliver a portion of their products in refillable packaging.

- Require large retailers to make products available in bulk.

CONEG did not adopt these recommendations.

Learning from Europe

The United States has the opportunity to learn much from Europe, where dramatic steps are being taken to reduce packaging waste. Strategies there are based on the principle that "the polluter pays," which makes producers of packaging responsible for the management of packaging waste. This, in effect, "internalizes" the cost of waste management and provides strong incentives for source reduction that will reduce the costs industry must bear for collecting and recycling packaging waste.

Germany's packaging act, passed in May 1991, virtually removes all packaging materials from the public waste stream. It also requires that bins be provided so consumers may leave secondary packaging (defined as packaging that promotes marketing or reduces theft) at retail stores, thereby providing a strong incentive for source reduction. It is expected that retailers will put pressure on manufacturers to deliver products without the extra layers that the retailers will otherwise have to collect.

While the German legislation focuses on recycling and does not set any source reduction goals, it does contain powerful incentives for source reduction that have already begun to reduce packaging waste. For example, Colgate-Palmolive has dropped the boxes for toothpaste tubes shipped into the German market, and Bristol-Myers has eliminated a plastic tray and cardboard box holding six deodorants together and is using only shrink wrap.[2]

In The Netherlands, a covenant signed by government and industry in June 1991 set goals of sending no packaging waste to landfills by 2000 and reducing generation of packaging waste to below the 1986 level by 2000.

The covenant contains specific source reduction initiatives for quantity reduction:

- No free bags at supermarkets as of July 1991
- No liquor to be sold in gift boxes or wrapping by the end of 1991
- Within one year, two-thirds of detergents to be concentrates
- End of small toothpaste tubes by September 1992
- Avoidance of multiple packaging and "excessive" packaging
- One guilder ($0.50) deposit on plastic bottles for water and soft drinks by October 1991
- A ban on advertising of beverages in containers that are not refillable (not in the covenant, but already announced)
- Stimulation of refill system for liquid detergents
- Wine to be sold in deposit bottles in 2000 shops, with some providing tap wine from a case

The covenant also includes specific initiatives for reducing toxic chemicals in the waste stream:

- End to lead caps on wine bottles (to be phased out in 3 to 4 years)
- End to heavy metals in inks and pigments
- End to beverage cartons bleached with chloride
- Reduction of polyvinyl chloride as a packaging material

The European Community (EC) has circulated a draft packaging directive for the Common Market countries that incorporates a "standstill principle," mandating that per capita generation of packaging waste in each member country in 2000 not exceed the 1990 EC average of 330 pounds per person per year. There has been opposition to this provision and it may not be included in the final directive.

Europe is also ahead of the United States in refilling beverage bottles. Denmark, in 1977, mandated that all beer and soda be sold in refillable bottles, and the new German law requires maintenance of the current refill rate of 17 percent for milk and 72 percent for all other beverages. Refill rates in the United States, in comparison, have dropped from 1960's levels of 46 percent for beer and 89 percent for soft drinks to below 10 percent. In Denmark and Germany, beer and soft drink bottles are refilled on average about 35 times.

Localities in Europe are also moving into action on packaging reduction. In Germany, the city of Munich has banned nonreturnable packaging for beverages and required that mineral water, beer, and milk be sold in bottles.[3] Aseptic milk containers and wax-coated cardboard will not be allowed. Possible sanctions include fines of up to DM 10,000 ($6000).

Canada's Bold Program

The Canadian Packaging Protocol adopted in 1989 set milestone targets to reduce packaging waste. By the end of 1992, packaging sent to disposal is to be no greater than 80 percent of the amount sent in 1988. By the end of 1996, this is to drop to 65 percent, and by the end of 2000, to 50 percent of the 1988 level. Half of these diversion goals are to be achieved by source reduction and the other half by recycling.

Possible Packaging Initiatives for Localities

Local solid waste plans often state that efforts to reduce packaging waste must be taken at the federal or state level and that the locality will support such efforts. Private companies prefer federal legislation on packaging standards and regulations. They fear "balkanization" of the packaging market since many products are marketed nationally and it is not practical to deal with different packaging regulations in different communities. It is also much more difficult for companies to monitor and influence packaging legislation in numerous states and localities, rather than at the federal level.

However, localities can do a lot more than just support efforts to reduce packaging waste by other governmental entities. They can work with state and federal officials to develop programs and legislation. They can address packaging directly by creating procurement policies, goals, bans, and economic incentives to reduce packaging waste. They can work cooperatively with companies and with community groups such as the League of Women Voters, which has already organized environmental shopping tours of supermarkets.

Many of the general source reduction strategies described in Chapters 6 to 13 can be used to specifically target packaging waste. Through procurement policy, for example, government, companies, and institutions can specify that minimal packaging be used and that suppliers ship in reusable containers or take their shipping containers back.

Impact of local packaging initiatives While local packaging legislation may not be an efficient policy approach, it can be effective. Packaging legislation in one large locality or state can reduce packaging waste throughout the country or even worldwide. For example, in response to a law in Iowa banning styrofoam blown with CFCs, American Telephone and Telegraph Company (AT&T) changed the packaging of its telephones worldwide. The company had been considering packaging changes, but the Iowa law was the catalyst for action. Since AT&T has a worldwide distribution system for telephones, it decided that producing a different package for the Iowa market would be very impractical and costly. Instead of using a large package with styrofoam and corrugated, AT&T now packages telephones in a smaller, folding cardboard box that is made of recycled material and is itself recyclable. The company is pleased with the new package, which reduces both waste and costs, but did not quantify these savings to INFORM.[4] This example illustrates the substantial leverage a local or state packaging law can have. A smaller locality would probably have to coordinate legislation with other localities in order to have such leverage.

Goals　A local solid waste plan could set a goal such as the EC "standstill principle" described above to cap the generation of packaging waste. In order to be meaningful, waste audits would be necessary to establish the baseline and to monitor progress. Goals could be set for packaging in general or for specific kinds of packaging such as plastic bags or corrugated. Goals could also be set for achieving a specified level of refillable beverage containers. Goals, of course, need to be supplemented with effective waste reduction programs.

Bans and incentives　While most bans in effect today focus on reducing toxicity or increasing recyclability, there are bans that could reduce the amount of packaging waste: bans on free bags at retail stores, bans on detergents not in concentrated form, bans on the sale of food in disposable packaging if it is consumed on premises, and bans on "egregious" packaging, to cite some examples. A ban on "egregious" packaging requires a clear definition, which might be "packages that have a package-to-product ratio exceeding 1" — in other words, packages that are greater in weight or volume than the product they contain.

Economic incentives can be used instead of bans. Taxes can be imposed on "egregious" packaging or on nonrefillable beverage containers. A "leave it in the store" policy, similar to the one in Germany, could, even on a voluntary basis, provide retailers with a strong incentive to carry the least overpackaged products. Variable waste disposal rates can provide an incentive to reduce packaging waste as long as there is a charge for recyclables as well as for trash.

Private sector initiatives　State and local governments can identify positive corporate packaging initiatives in their areas and aid in their replication through technical assistance, grant, and awards programs, as discussed in Chapter 7. Some of the source reduction initiatives adopted by private companies that were described in that chapter involve strategies to reduce packaging waste. The automobile industry, for example, has put pressure on its suppliers to ship all products in reusable/returnable shipping containers. The Herman Miller furniture company's switch to reusable blankets (Chapter 9) for shipping furniture was in part a response to customers who did not want to have to dispose of a lot of plastic and corrugated packaging. INFORM research on less wasteful products and packaging will, in forthcoming reports, identify private sector initiatives that reduce waste, including case studies of companies that are refilling bottles and shifting to reusable shipping containers.

Working with companies The CONEG model can be replicated in localities. Councils consisting of representatives from government, private companies, and environmental organizations can be formed to recommend action to be taken on the local, state, and federal level to reduce packaging waste.

In 1990, Indiana established a state-level packaging Waste Reduction Task Force composed of representatives from the packaging and manufacturing industries, the Indiana Department of Environmental Management, and environmental and consumer groups. Like CONEG, this group issued guidelines and recommendations to reduce packaging waste, but failed to set a separate goal for source reduction. The task force has recommended the establishment of a body to monitor packaging and report annually to the governor and state legislators. The objectives of the group would include developing specific quantifiable goals for source reduction and recycling, and educating the public on packaging waste reduction.

Local pilot programs can provide valuable information on which state and federal programs may be based. Measuring the amount of waste reduction produced by specific initiatives is more easily achieved at the local level and can be used to demonstrate the efficacy of waste reduction programs. New York City's "Waste Prevention Partnership," for example, includes supermarkets, small food retailers, and dry cleaners (see Chapter 8). It has the potential for developing waste reduction programs that can be implemented by government and the private sector. As part of this program, one supermarket chain, D'Agostino's, agreed not to double bag groceries unless the customer requested it and will soon credit customers 2 cents for not taking a bag. This effort to reduce packaging waste cut expenditures for purchasing bags in half (see Chapter 8), despite an increase in grocery sales.

Notes

1. Personal communication, Carol Bell, Rhode Island Department of Environmental Management, Division of Environmental Coordination, to Caroline Gelb, INFORM November 13, 1991.

2. Environmental Data Services, Ltd., *The Ends Report,* October 1991.

3. *Warmer Bulletin,* November 1991, p. 13.

4. Presentation by Barry Dambach, AT&T, at Environmental Forum 1991, Chattanooga, TN, April 26, 1991.

Chapter 13 Reducing Toxic Materials in Solid Waste

by Nancy Lilienthal

From paints, batteries, and home pesticides to cleaning products, fingernail polish, and motor oil, a variety of consumer products contain synthetic and other chemicals that are known or suspected to be harmful to our health or the environment. When such products, or the containers they come in, are disposed of, these chemicals make their way to landfills, garbage incinerators, home septic systems, and public sewer systems. From there, they can move into the air, land, and water, making the reduction of the amount of toxic chemicals in the municipal solid waste stream an important focus of source reduction attention.

Over the last half-century, chemically based products have become more and more a part of our lives. Annual production of synthetic organic chemicals has increased 15-fold since World War II,[1] to a total of 267 billion pounds in 1989.[2] Some 70,000 chemicals are now in use worldwide,[3] and the US Environmental Protection Agency (EPA) reviews, on average, 1000 new chemicals each year that are proposed for manufacture.[4]

As our chemical use has grown, so has public concern about the environmental and health risks from chemicals that may be "toxic," i.e., that may (given sufficient exposure) cause serious health effects such as poisoning, cancer, or birth defects. Yet these risks are still poorly understood. A National Research Council report estimated that less than 30 percent of the more than 17,000 chemicals used in pesticides, cosmetics, food, and drugs have been tested to the point where their ability to cause cancer or reproductive damage is understood. Less than 10 percent of other chemicals in use have been so tested.[5]

The word "toxic" used in this report refers to chemicals on the list EPA uses for its annual industrial Toxics Release Inventory (TRI). The list, based on scientific toxicity data, includes over 300 chemicals and 20 classes of chemicals. Some 23,000 large US generators of manufacturing wastes are required to report annually to the EPA their releases of these chemicals to the air, land, water, and off-site treatment and disposal facilities. TRI chemicals represent a wide range of toxicities.

❖ Toxics in the Municipal Waste Stream

Most products in the municipal solid waste stream that contain toxic chemicals are referred to as "household hazardous waste" (HHW). This includes paints, used motor oil, certain cleaning products, batteries, pesticides, and many other products. No exact figures are available as to how much HHW is generated. However, a 1987 EPA study of two localities, conducted by the University of Arizona, put the amount at between 0.35 and 0.45 percent of the waste stream by weight.[6] Assuming per capita total waste generation of 4-5 pounds a day, this translates into an average of 5-8 pounds per person per year going to waste disposal facilities. The US Congress's Office of Technology Assessment (OTA) believes this estimate is probably low, since it includes only that "portion of the waste that contains the hazardous ingredients, not including contaminated containers or other contaminated articles such as paint brushes and oil-soaked rags." However, OTA concludes from these studies that, in general, HHW comprises less than 1 percent by weight of the municipal waste stream.[7]

Assuming conservatively that HHW accounts for 0.4 percent of a US waste stream of 180 million tons a year, HHW would amount to some 720,000 tons a year. While this may seem insignificant compared to the millions of tons of toxic solid waste generated by industry, industrial toxic wastes are strictly regulated and must be disposed of in special landfills and incinerators designed for toxic and other hazardous waste. These disposal facilities are all subject to strict government regulations. Municipal landfills and incinerators, on the other hand, are not designed to receive primarily hazardous wastes and are not as strictly regulated.

While HHW is the most important *household* source of toxics in the municipal waste stream, it is not the only source. Some portion of hazardous materials in the municipal waste stream comes from businesses that generate small amounts of hazardous waste (less than 100 kilograms, or 220 pounds, a month), such as vehicle maintenance shops, which are allowed by law to send their wastes to municipal solid waste landfills and incinerators.[8] The amount of hazardous waste of this type going to municipal disposal facilities is unknown. Additionally, paper and plastics, not commonly considered HHW, appear to be a major source of chlorine in the waste stream.[9] When incinerated, chlorine compounds can form dioxin, a substance suspected to be highly toxic. Plastics and inks used in packaging are sources of cadmium and other toxic metals.

❖ Pollution Caused by the Disposal of Products Containing Toxic Chemicals

Toxic chemicals in products can enter the environment when the products are disposed of in landfills, incinerated, and even recycled. The EPA and others have measured the concentrations of toxic pollutants discharged by waste disposal facilities. However, the actual health and environmental risks associated with this pollution are still in many cases a matter of scientific debate.[10] More information characterizing risks is needed. Nevertheless, source reduction of toxics in products is the primary strategic option that prevents toxics from entering the municipal waste stream.

When products containing toxic chemicals are dumped into a landfill, their toxic constituents may drain or leach into acidic liquids in the landfill (leachate) which can then flow from the landfill into groundwater or surface waters. More than 100 potentially harmful substances have been identified in landfill leachate.[11] Many, such as lead, methylene chloride, and vinyl chloride, have been found at concentrations far exceeding federal regulatory standards.[12] Only 11 percent of existing landfills have any type of leachate collection system to minimize water pollution.[13]

In addition to threatening groundwater and surface water, toxic chemicals in landfills can generate toxic emissions that may pose risks to human health. Chemicals measured in landfill emissions include toluene, vinyl chloride, and xylenes.[14]

The presence of toxic chemicals in garbage incinerator emissions and ash, caused largely by the incineration of products containing toxic chemicals or pollutant precursors, has been widely documented. For example, toxic heavy metals (such as lead, mercury, and cadmium) can be emitted into the air, as can dioxins and furans and other toxic organic chemicals formed during the combustion of chlorine and chlorinated substances.[15] The amounts of such substances actually emitted depend on a variety of factors, including incinerator design and operation as well as the amounts of precursor chemicals in the incinerator feedstock. Toxic chemicals have also been found in incinerator ash. If the ash blows around during handling, it can carry these chemicals into the air and water; also, toxic substances in the ash can leach out of landfills where ash is disposed of.

Certain processes used to recycle products containing toxic chemicals, such as paper deinking, may create toxic wastes. In addition, sanitation workers who collect and handle garbage may be exposed to a variety of hazardous chemicals from household hazardous wastes. A study conducted in Los Angeles between 1980 and 1985 showed 36 injuries due to identified chemical burns, 29 due to inhaled chemicals, and 93 from unidentified chemicals.[16]

❖ Chemicals and Products of Greatest Concern

Chemicals

INFORM research indicates that at least 40 Toxics Release Inventory chemicals are used in household products, and the number may be much larger.[17] (INFORM has done no independent research or product testing to determine the chemical content of products.) Further research still needs to be done to identify and quantify specific toxic substances in municipal solid waste and to identify more precisely the major sources of these substances. EPA's Office of Solid Waste has begun such research by looking at three major categories of toxic substances in the waste stream:

- metals, such as lead, mercury, and cadmium
- chlorinated organics (organic, or carbon-based, chemicals containing chlorine) such as methylene chloride, perchloroethylene, and the chlorinated compounds in bleached paper
- aromatic solvents (solvents containing organic chemicals based on the benzene ring), such as toluene and xylenes

EPA's next steps are to narrow its focus to a few specific chemicals within these categories and to identify the major sources of those chemicals in the waste stream. However, some information is already available on sources of these important classes of toxic chemicals.

Less is known about the sources of chlorinated compounds in the waste stream than is true for the sources of metals. However, paper bleached with chlorine (most paper) and plastics such as polyvinyl chloride are generally recognized as major sources of chlorine in the waste stream. A 1985 National Bureau of Standards study estimated that paper contributed 56 percent of the chlorine in the combustible portion of the Baltimore, Maryland waste stream, and that plastics contributed 52 percent in Brooklyn, New York .[18] Chlorinated solvents in paints, paint strippers, and some cleaning products are also sources of chlorine. The incineration of chlorine, as mentioned above, can create dioxin in incinerator emissions and ash.

Products

The 1987 study conducted by the University of Arizona for the EPA estimated the annual generation rates of various types of HHW in New Orleans, Louisiana, and Marin County, California. The study found that just three items — motor oil, paint, and batteries — accounted for 45-50 percent of HHW by weight in both locations. The largest general categories of HHW were: automotive maintenance (12-21 percent), batteries and electrical devices (12-27 percent), household cleaners (13-15 percent), household maintenance (paint, glue, varnish, etc.) (28-43 percent), pesticides and yard maintenance (1-9 percent), and cosmetics (4-5 percent).[19] The composition of HHW, as with the waste stream generally, varies from location to location.

Tables 13-1 and **13-2** list examples of Toxics Release Inventory (TRI) chemicals that may be contained in products, and a range of products that do or may contain TRI chemicals. The tables show only the TRI chemicals mentioned by the listed sources. Products may also contain other TRI chemicals not shown in the tables.

The information in these tables, including which chemicals may be in which types of products and possible alternatives to using the products, is not comprehensive. It comes from the government and nongovernment sources cited at the end of this section.

Product manufacturers consider the exact chemical content of their individual brands of products proprietary in many cases, and it is impossible to determine this without laboratory testing of the products. Thus, not all product brands in a product category listed here necessarily contain the chemicals shown. Some product formulations may have changed since the information sources used for the tables were published. However, the tables do indicate which types of products might be priorities in municipal programs to reduce toxic chemicals in the waste stream.

Table 13-1 lists 39 TRI chemicals likely to be found in common products, along with the chemicals' potential effects as identified by the US Environmental Protection Agency, and some of the products in which these chemicals may be found. The information on potential health effects of the chemicals in Table 13-1 comes from a toxicity matrix developed by the EPA from a preliminary search of existing scientific databases.

Table 13-2 lists 38 of the key types of products of concern and identifies some TRI chemicals they do or may contain. The table also provides some suggestions as to less toxic or nontoxic alternatives to, or substitutes for, these products, although these alternatives may not in all cases be as effective or convenient as their counterparts which may contain TRI chemicals.

The principal sources of information used in Tables 13-1 and 13-2 to identify which chemicals may be found in which products are:

- Chemical Specialties Manufacturers Association, telephone consultation.

- Dadd, Debra Lynn. *Nontoxic, Natural and Earthwise.* Jeremy Tarcher, 1990.

- Franklin Associates, Ltd. *Characterization of Products Containing Lead and Cadmium in Municipal Solid Waste in the United States, 1970-2000.*

- The Garbage Project, University of Arizona. *Hazardous Constituents of Common Household Products.* 1989.

- Levin, Hal. "Building Materials and Indoor Air Quality," *Occupational Medicine: State of the Art Reviews.* October-December, 1989.

- US Environmental Protection Agency. *Household Solvent Products: A Shelf Survey with Laboratory Analysis.* 1987.

- US Environmental Protection Agency. *EPA Indoor Air Quality Implementation Plan. 1987.*

- US Environmental Protection Agency. *A Survey of Household Hazardous Wastes and Related Collection Programs.* 1986.

- US Environmental Protection Agency. *Solid Waste Disposal in the United States*. 1988.

- US Food and Drug Administration. *FDA's Cosmetics Handbook*. 1991.

The principal sources of suggestions as to less toxic or nontoxic alternatives include:

- Berthold-Bond, Annie. *Clean and Green*. Ceres Press, 1990.

- Dadd, Debra Lynn. *Nontoxic, Natural and Earthwise*. Jeremy Tarcher, 1990.

- Greenpeace. *Stepping Lightly on the Earth: Everyone's Guide to Toxics in the Home*. 1990.

- Clean Water Fund and New Jersey Environmental Federation. "Home SAFE Home" project.

- Washington Toxics Coalition. *A Database of Safer Substitutes for Hazardous Household Products, Phase One Report*. 1991.

Further discussion of the material in these tables and the issues related to toxic chemicals in consumer and building products can be found in INFORM's 1992 publication, *Tackling Toxics in Everyday Products: A Directory of Organizations*.

Table 13-1 *TRI Chemicals Likely to Be Found in Common Products*

TRI Chemical	Potential Health Effects*	Some of the Products in Which It May Be Found
Metals		
Cadmium	A,C,Ch,D,E,R	Nickel/cadmium (rechargeable) batteries account for an estimated 54% of cadmium in the waste stream after recycling; plastics account for 28%; electronics, 9%; appliances, 5%; pigments (in plastics, paints, and inks), 4%. Also metal coatings and platings.
Chromium (hexavalent form)	C,Ch,E	Inks, pigments
Lead	D,N,R	Automobile batteries account for an estimated 65% of lead in the waste stream after recycling; consumer electronics for 27%; glass, 4%; plastics, 2%; and other, 2%, including solder in steel cans, printing and packaging inks, the base of incandescent light bulbs, wine bottle caps, stains and varnishes, and hair dyes.

continued

Table 13-1 *TRI Chemicals Likely to Be Found in Common Products (continued)*

TRI Chemical	Potential Health Effects*	Some of the Products in Which It May Be Found
Mercury	Ch,E,N,R	Household batteries, fluorescent light bulbs, switches for lights and thermostats, thermometers
Nickel	C,Ch,D,R	Nickel-cadmium batteries
Zinc	E	Household batteries
Chlorinated Organics		
Chlorine	A,Ch,E	Paper products and plastics, principally
Chlorothalonil	C,E	Lawn care chemicals
Hydrochloric acid	A,Ch	Toilet cleaners
Methylene chloride (dichloromethane)	C	Spray paints, rust paints, paint strippers, adhesives, pesticides
Paradichlorobenzene (1,4-dichlorobenzene)	C,Ch,E	Mothballs/crystals, certain air fresheners and toilet deodorizers
Pentachlorophenol	A,D,E,R	Varnishes/stains/sealants
Perchloroethylene (tetrachloroethylene)	C,Ch,D,E,R	Rug/upholstery cleaners, spot removers
1,1,1-Trichloroethane (Methyl chloroform)	D,E,R	Dry cleaning, spot removers, fabrics, typewriter correction fluid, adhesives/glues
Vinyl chloride	C,Ch,D,M,R	A variety of household plastics, including furnishings and apparel
Aromatic Compounds		
Ethylbenzene	Ch,D,E,R	Paint
Naphthalene	E	Mothballs, adhesives, pesticides
Toluene	D,E,M,R	Paints and paint thinners, furniture strippers, adhesives, art supplies, nail polish
Xylenes	Ch,D,E,R	Paints, adhesives, pesticides, art supplies, furniture strippers
Other TRI Chemicals		
Acetaldehyde	A,C,E	Adhesives
Acetone	Ch,E	Varnishes, lacquers, paint thinners, furniture strippers, adhesives, nail polish remover, art supplies
Acrylonitrile	A,C,D,E,R	Fabrics, apparel
Ammonia	A,CH,E	All-purpose cleaners, glass cleaners, hair dyes

continued

Table 13-1 *TRI Chemicals Likely to Be Found in Common Products (continued)*

TRI Chemical	Potential Health Effects*	Some of the Products in Which It May Be Found
Carbaryl	A,Ch,E,N,R	Pesticides, pet flea and tick treatments
Cresol	A,Ch,E	Art supplies, disinfectants
2,4-D	A,C,D,E,R	Lawn care chemicals
Dibutyl phthalate	Ch,D,E,R	Paint, adhesives
Di(2-ethylhexyl) phthalate (DEHP)	C,Ch,D,E,M,R	Pliable plastics, fabrics, apparel, hair sprays
Diethyl phthalate	A,E	Paints, adhesives
2-Ethoxyethanol	Ch,D,R	Polyurethane wood finishes
Ethylene glycol	Ch	Deodorants/antiperspirants, paints
Formaldehyde	C,Ch,E,M,R	Plywood, particle board, clothing, adhesives, upholstery fabrics, nail polish
Methanol	N	Paint thinners and strippers, adhesives
Methyl ethyl ketone	Ch,D,N,R	Paint thinners, adhesives, cleaners, waxes
Methyl isobutyl ketone	Ch,N	Paint thinners, pesticides
n-Butyl alcohol	Ch	Paint strippers, perfumes, aftershave lotions
Phenol	A,D,E	Art supplies, adhesives
Phosphoric acid	No health effects listed.	Metal polishes
Sulfuric acid	A,Ch,E	Automotive (lead-acid) batteries

* Information on potential health effects comes from the US EPA's toxicity matrix of chemicals (published by several environmental organizations) which the EPA developed from a preliminary survey of existing scientific databases. If a particular health effect is not shown after a chemical, it may mean that this health effect is not a concern for that chemical, or that the effect was not reported, or that the chemical has not been tested.

Key to Abbreviations:
A, Acute toxin (may cause harm through short exposure; e.g., burns, breathing problems)
C, Carcinogen (may cause cancer)
Ch, Chronic toxin (may cause damage over a period of years)
D, Developmental effects (may cause miscarriages or birth defects)
E, Environmental toxin (may harm wildlife)
M, Mutagen (may affect genes and chromosomes)
N, Neurotoxin (may harm nervous system: brain, nerves, spinal chord)
R, Reproductive effects (may effect male or female fertility)

Table 13-2 *Key Products Likely to Contain TRI Chemicals and Less Toxic or Nontoxic Alternatives*

Products of Concern	TRI Chemicals They May Contain	Suggestions as to Less Toxic or Nontoxic Alternatives
Automotive Maintenance		
Antifreeze/coolant	Ethylene glycol, methanol	NA*
Engine treatment (transmisson and oil additives, fuel additives, carburetor cleaners)	Methylene chloride, toluene, 1,1,2-trichloroethane, xylenes	NA
Oil and transmission fluid	Lead	NA
Other automotive (solvents, etc.)	Toluene	NA
Batteries and Electrical		
Automotive batteries	Lead, sulfuric acid	NA
Household batteries	Cadmium, mercury, nickel, zinc	Where possible, use manual rather than battery-run items. If batteries are necessary, using rechargeables means fewer are used.
Household Cleaners		
Air fresheners	Paradichlorobenzene (in those that are in form of white blocks/crystals and those used to deodorize toilets)	Keep areas clean. Sprinkle baking soda in trash cans, or baking soda or white vinegar in open dishes. Use herbal potpourri for bathrooms.
All-purpose cleaners	Ammonia	One teaspoon liquid soap and/or borax per quart warm or hot water, or 1/2 cup washing soda per bucket of water. Add a little lemon juice or vinegar to cut grease.
Chlorine bleach and scouring powder	Chlorine	Powdered, nonchlorine bleach.
Cleaners/waxes	Methyl ethyl ketone	Nontoxic cleaners and waxes.
Disinfectants	Chlorine, cresol	Borax (1/2 cup in gallon of hot water).
Glass cleaners	Ammonia	Half white vinegar, half water. (May first need to use rubbing alcohol to remove residues of commercial cleaners.) When dry, rub with newspaper to avoid streaks.
Metal polishes	Acetone, phosphoric acid	Boil silver flatware in water with baking soda and salt. Polish silver and stainless flatware with paste of baking soda and water. For brass, equal parts salt and flour with a little vinegar. For copper, lemon juice and salt. For chrome, white flour in a dry rag.

* NA, no commercially available alternatives.

continued

Products of Concern	TRI Chemicals They May Contain	Suggestions as to Less Toxic or Nontoxic Alternatives
Rug/upholstery cleaners	Perchloroethylene (in dry-cleaning types, not water-based shampoos)	For removing odors and fresh grime, sprinkle cornstarch and borax, or baking soda, liberally on surface. Let sit 1/2 hour, then vacuum. For spots/stains, see "spot removers," below. Buy dark-color carpets.
Spot removers	Perchloroethylene, 1,1,1-trichloroethane	Variety of homemade alternatives, depending on type of spot or stain. For coffee or chocolate, use borax and water. For ketchup, gravy, or red wine, use club soda. Clean stains right away.
Toilet cleaners	Hydrochloric acid	Use baking soda. If badly stained, use a paste of lemon juice and borax.
Household Maintenance		
Adhesives, glues	Acetaldehyde, acetone, formaldehyde, dibutyl and diethyl phthalates, methanol, methyl ethyl ketone, methylene chloride, 1,1,1-trichloroethane, naphthalene, phenol, toluene, xylenes	Avoid solvent-based adhesives when possible. Less toxic adhesives available. For light jobs such as paper, use stick-type or basic white glue.
Furniture strippers	Acetone, methyl ethyl ketone, methylene chloride, toluene, xylenes	Less toxic, solvent-free strippers available.
Paints	Dibutyl and diethyl phthalate, ethylbenzene, ethylene glycol, methylene chloride, toluene, xylene	Less toxic and nontoxic paints available. Latex (water-based) paints are less toxic than oil-based paints.
Paint strippers/ removers	Acetone, methanol, methylene chloride, n-butyl alcohol, toluene	Use water-based paints that can be removed with water. Or, wearing gloves, mix 1 pound trisodium phosphate in 1 gallon of water. Brush onto surface and let sit for $1/2$ hour. Scrape off softened paint.
Paint thinners	Acetone, methanol, methyl ethyl ketone, methyl isobutyl ketone, toluene	Use water-based paints that do not need thinner.
Polyurethane wood finishes	2-Ethoxyethanol	Less toxic finishes available.

continued

Products of Concern	TRI Chemicals They May Contain	Suggestions as to Less Toxic or Nontoxic Alternatives
Varnishes/stains/ sealants	Acetone, lead, methanol, pentachlorophenol	Nontoxic or less toxic brands available.
Pesticides and Yard Maintenance Lawn care chemicals	Chlorothalonil, 2,4-D	Consult sources of information on organic gardening.
Pesticides	Carbaryl, methyl iso-butyl ketone, methylene chloride, naphthalene, 1,1,1-trichloroethane, xylene	For houseplants, use dishwashing or bar soap and water. For roaches, boric acid (keep away from children). For ants, equal amounts of sugar and borax, or cayenne pepper. For termites, use least toxic pesticide effective and only in affected areas. (To avoid termites, keep wood structures away from contact with soil and keep firewood and wood scraps away from house.) In garden, 2 tablespoons of liquid pure soap in 1 quart water, or insecticidal soap. Other nontoxic alternatives and products available.
Pet flea and tick treatments	Carbaryl	Rub pet's coat with brewer's yeast and add yeast to pet food. Wash pet with warm, soapy water and rinse with 1/2 cup dried or fresh rosemary in 1 quart water. (Boil, then steep 20 minutes, strain and cool.) Spray or sponge onto pet and allow to dry (do not use towel). Insecticidal soaps or pet shampoo containing d-limonene available in pet shops.
Personal Care/Cosmetics Deodorants/antiperspirants	Ethylene glycol	Nontoxic brands, or baking soda rubbed on.
Fingernail polishes	Formaldehyde, toluene	Rub paste of powdered, dried henna (powdered tree leaves available in several shades) and water into nails, and buff.
Fingernail polish removers	Acetone	See substitute for nail polish, above; remover not needed.
Hair dyes	Ammonia, lead	Rinse with pure henna (powdered tree leaves available in several shades). Do not use on face, brows, or on dyed or to-be-permed hair. Or, rinse with lemon juice in warm water and sun-dry hair to lighten. To darken, rinse with strong black coffee or tea.

continued

Table 13-2 *Key Products Likely to Contain TRI Chemicals and Less Toxic or Nontoxic Alternatives (continued)*

Products of Concern	TRI Chemicals They May Contain	Suggestions as to Less Toxic or Nontoxic Alternatives
Hair sprays	Di(2-ethylhexyl) phthalate (DEHP)	Chop 1 lemon and boil in 2 cups water until volume reduced by half. Cool, strain, then refrigerate in spray bottle.
Perfumes, aftershave lotions	n-Butyl alcohol	Natural essential oils.
Other products		
Art supplies	Acetone, cresol, phenol, toluene, xylene	Water-based paints, inks, shellacs, felt-tip markers.
Fabrics/apparel	Acrylonitrile, di(2-ethylhexyl) phthalate (DEHP), formaldehyde, polyvinyl chloride, 1,1,1-trichloro-ethane, vinyl chloride	Untreated fabrics available. Avoid plastics.
Paper	Chlorinated compounds, dioxins	Use unbleached paper products when possible.
Plastics	Chlorinated compounds, cadmium, lead	Avoid use of disposable plastic items when possible.
Printing and packaging inks	Cadmium, chromium, lead	Request manufacturers/printers to use inks that do not contain these metals.
Typewriter correction fluid	1,1,1-Trichloroethane	Correction tape or tape that lifts off errors.

❖ Establishing a Program for Reducing Toxics in Municipal Solid Waste

There are two basic strategies for reducing the amount of toxic chemicals in the waste stream. The first and most important method is to encourage the purchase and use of products containing less toxic constituents than products commonly in use — for example, nickel-metal hydride instead of nickel-cadmium batteries, or paint with reduced or no toxic solvents and pigments.

The second basic strategy is to encourage people who already have products containing toxic chemicals in their homes or businesses to use them up or give them away, instead of throwing them out. This second method may keep toxic substances out of the municipal waste stream, but will increase the use of the products, which may pose problems of its own. For example, the use of many products containing toxic chemicals has been found to cause such environmental problems as indoor air pollution and smog.

Household Hazardous Waste Collection: Not Source Reduction

A major response to the problem of HHW has been the organizing by government or private groups of household hazardous waste collection days. While such HHW collection programs have been very important in keeping products containing toxic chemicals that have accumulated in homes or offices out of disposal facilities, they are not source reduction. The HHW must be collected and disposed of in some way.

Furthermore, to date these programs have been very expensive. According to the EPA, participation in such programs is typically quite low (less than 1 percent of households in the community), making the costs per pound of waste collected and disposed of often very high — more than $9.00 a pound, or $18,000 a ton.[20] In comparison, the cost of solid waste collection in the Northeast is about $65 a ton, and the average national disposal tip fee is about $27 a ton, making typical costs of solid waste collection and disposal about $100 a ton.[21]

The most economical and environmentally beneficial strategy is for homes and offices not to generate such wastes in the first place. This can be done only through reducing the use of toxics at the source. HHW collection programs are indeed valuable as a source reduction tool to the extent that they educate people as to what types of products to avoid and what substitutes are available.

Setting Source Reduction Goals for Toxic Chemicals

Setting source reduction goals for toxic chemicals is not as straightforward as setting goals for reducing the volume of garbage because it is not possible to simply set a percentage reduction goal for toxics in the waste stream. First, there are no accurate data as to the overall amount of toxic materials in municipal waste streams. Second, the toxicities and environmental impacts of substances in the waste stream vary widely, so setting the same percentage reduction goal for all chemicals does not make sense. A pound of mercury in an incinerator, for example, would have very different environmental impacts from a pound of hydrochloric acid.

It is also difficult to set source reduction goals for individual toxic chemicals in the waste stream because, except for certain well-studied chemicals like mercury and cadmium, there are no reliable data on the amounts of individual chemicals in a municipal waste stream, all the significant product sources of each particular chemical, or the amount of each chemical different products may be contributing.

The most practical way to set toxicity source reduction goals is to set goals for reducing the purchase and disposal of specific products that are known to contain particular toxic substances. For example, a community could set a goal of reducing the sales or disposal of paint strippers containing methylene chloride by a certain percentage over a five-year period. Meeting such a goal would almost certainly reduce methylene chloride in the waste stream, but it would not be possible to say by how much, because of a lack of reliable data on how much there was to begin with or on how much methylene chloride paint strippers had been contributing relative to other sources.

In some cases, especially for metals such as lead and cadmium, national estimates exist for the percentage of a certain toxic chemical in the waste stream contributed by a specific category of product (see Table 13-1). In such cases, it is possible to roughly estimate the amount by which the chemical will be reduced in the waste stream if use or disposal of the product is reduced by a certain amount (acknowledging the fact that every community's waste composition is different). For example, nickel-cadmium batteries are estimated to account for 54 percent of cadmium in the waste stream nationwide, after recycling.[22] Therefore, if a municipality could reduce sales or disposal of nickel-cadmium batteries by 50 percent, cadmium in the waste stream should be reduced, over time, by roughly 27 percent (54% × 50%).

Given these difficulties, there are three basic steps to setting goals for reducing toxic chemicals in the municipal waste stream: identifying chemicals and products of concern, identifying existing less toxic alternatives, and setting reduction goals for individual products.

Step 1: Identifying chemicals and products of concern. Information on many of the chemicals and products of concern in the waste stream, useful for targeting municipal programs, appears in Tables 13-1 and 13-2. For some types of products listed in the tables, such as batteries, the major toxic constituents are well known. For other products, less information is available, and many product formulations have changed over time. In addition, different brand names within a product category (such as latex paint) may contain different chemicals.

More information on the specific toxic substances contained in various categories of products is being compiled by various US Environmental Protection Agency offices concerned about the effect toxics in products have on indoor air quality and smog formation, as well as those concerned about toxics in municipal solid waste.

Step 2: Identifying existing less toxic alternatives. After identifying substances and products on which to focus, the next step is to obtain information on less toxic alternatives that exist for each type of product. Such alternatives may be similar products formulated with less toxic chemicals (e.g., paint strippers without methylene chloride) or different types of products (e.g., electrically or hand-operated gadgets and appliances instead of battery-run ones).

A knowledge of realistic alternatives is the key to setting realistic reduction goals. For example, paint strippers without methylene chloride or large amounts of other toxic chemicals are available, so an ambitious goal could be set for reducing sales of the more toxic paint strippers, or for restricting their disposal. However, there is currently no commercially available automobile battery that does not contain lead, so a goal could not be set for reducing sales of this product (although its disposal could be reduced through recycling).

In assessing alternatives to a specific product, it is important to watch out for alternative products that may have eliminated one chemical of concern but substituted another. For example, in the case of paint strippers, some have substituted toluene for methylene chloride. However, others have significantly reduced the toxic content overall.

Other important points to consider in assessing alternatives are their cost and their effectiveness/convenience compared to the products to be reduced. If an alternative is much more costly and/or much less effective than the product in use, there is likely to be consumer resistance to buying and using it unless some type of subsidy is offered or waste tax created.

Some general, summary information on less toxic alternatives is provided in Table 13-2. For more information, see the sources of information for that table listed on pages 136-139.

Step 3: Setting toxic reduction goals for each product. In setting reduction goals for each product of concern, municipal officials must keep in mind realistic alternatives to the products to be reduced.

Measuring Progress in Reducing Toxics in the Waste Stream

As with goals for reducing the amount of garbage, goals for reducing the sale and disposal of products contributing to waste stream toxicity should be measurable. There are several ways to do this: measuring both progress in reducing the sale and/ or disposal of products targeted for reduction, and success in keeping these products out of the waste stream. Such measurement tactics include:

- Monitoring the sales of targeted products in the municipality.

- Monitoring activity at any product exchange centers that may be established, to develop information on how much of unused targeted products (such as unused paint) is being acquired and (presumably) used up.

- Sampling the waste stream for the targeted products. This would provide information on the waste stream impact of measures to reduce the sale and disposal of these products. (The results would need to take into account the effect of any HHW collection programs.)

Changes in population and employment also need to be taken into account in assessing measurements of reductions in toxics in the waste stream, as they do for measurements of reductions in the amount of waste.

❖ Strategies for Achieving Source Reduction of Toxics

Strategies for achieving reductions in the amounts of targeted products containing toxic chemicals that are used and need to be disposed of can be aimed at purchasers and users of the products and at manufacturers. Purchasers and users of the products include consumers, retailers, other commercial enterprises, and government and other institutions. Efforts here can be directed at reducing the purchase of the targeted products and encouraging people to use up or give away targeted products stored in their homes and offices. For product manufacturers, the aim is to encourage them to reformulate or redesign targeted products to reduce or eliminate their toxic components.

Strategies for reducing toxics in the waste stream fall into the same seven general categories identified for reducing the quantity of the waste stream: (1) government source reduction programs, including procurement; (2) institutional source reduction programs; (3) government assistance programs, including reuse programs; (4) business source reduction programs; (5) public education (consumers and schools); (6) economic incentives and disincentives; and (7) regulatory measures, including labeling requirements and product bans.

❖ Government Procurement and Operations

Changes in government procurement policies can reduce the toxicity as well as the quantity of waste and can encourage manufacturers to reformulate products to reduce toxicity. For example, government can specify unbleached paper for many uses, less toxic cleaning supplies, and less toxic pesticides. It can purchase mechanically and solar-operated (rather than battery-operated) products, such as pencil sharpeners and calculators. When batteries are needed, it can favor use of rechargeable batteries. It can require that its materials be printed with nontoxic inks.

Changes in government operations can also reduce the amount of toxics in the waste stream. For example, latex paint can often be used instead of oil-based paint. In addition, paint can be used more efficiently by mixing the right amount so as to waste less. Leftover paint can be remixed or given to stores that can mix them (as discussed under "Reuse Programs," below).

❖ Institutional Source Reduction Programs

Broward County, Florida, Hospitals

Regulations at Broward County, Florida, waste-to-energy incinerators prohibit the disposal of batteries and other wastes containing mercury. To help comply with these regulations, Broward County's Resource Recovery Office asked hospitals in the county to find ways to keep mercury from medical batteries and other hospital supplies out of the waste stream.[23] Discussions between the county and some 16 hospitals to review the types, uses, and quantities of batteries the hospitals were using confirmed that 8.4-volt mercury batteries were being widely used in cardiac monitors and in intermediate care units. The hospitals were disposing of these batteries in regular or biohazardous waste containers. The county estimated that more than 63,000 of these batteries, containing nearly a ton of mercury, were entering the waste stream each year. The county found that a county-funded program to recycle medical batteries would be prohibitively expensive.

One of the hospitals suggested the substitution of a zinc-air battery. When administrators of hospitals using mercury batteries focused on the problem of mercury-containing batteries and learned that a viable alternative was available, most made the switch to zinc-air. In addition, they found that making the switch saved money. While zinc-air batteries are typically more expensive than mercury batteries, they last longer, requiring less frequent replacement and disposal. Further, the hospitals avoid the costs of disposing of the mercury batteries as more expensive hazardous waste.

❖ Government Programs to Stimulate Source Reduction

Reuse Programs

Government can sponsor and publicize reuse programs that keep toxics out of the waste stream in addition to reducing the amount of garbage. Two examples of such programs are True Colors Home Decorating in Montpelier, Vermont, and Green Alternatives, Inc., in San Jose, California. New York City's Materials for the Arts program, described in Chapter 8 under "Reuse Programs," is another example.

■
True Colors Home Decorating
Contact
Bill McQuiggan
Address
114 River St.,
Montpelier, VT 05602
Phone
802-223-1616
■

Reclaimed Paint: True Colors Home Decorating, Montpelier, Vermont, and Green Alternatives, Inc., San Jose, California Remixing leftover latex paints provides an opportunity for reducing the amount of paint in the waste stream, while making less expensive paint available to consumers. True Colors Home Decorating, a small home decorating supply store in Montpelier, Vermont, and Green Alternatives, Inc., in San Jose, California, both sell reclaimed latex paint. These two examples demonstrate that a program to reuse paint can be developed on either a small or large scale. Government can be involved in the development of this type of program and/or use the end product as part of a source reduction procurement program.

True Colors Home Decorating takes leftover paint from donors, remixes it, and resells it. The cans sell for $5 a gallon (less than one-third the price of a gallon of comparable new paint), or $19.99 for a five-gallon bucket. Only latex paint is accepted, and it must be less than two years old to limit the risk of making paint that is not equal to the quality of new paint.

The program was co-designed by the store's owner and the Central Vermont Regional Planning Commission.[24] Between the start of the program in January

1990 and mid-1991, approximately 200 gallons of paint were remixed. Costs of running the program were not recouped from the sale of reclaimed paint during that period since there were additional labor costs for handling the material. Half of the remixed paint was sold and the other half was donated to local groups, such as a school and a day care center.[25]

■
Green Alternatives, Inc. ·
Contact
Tobi Romero
Address
P.O. Box 4190, Santa Clara, CA 95056
Phone
408-441-0241
■

Green Alternatives, Inc., started collecting, processing, and selling reclaimed latex paint in 1988. Between then and early 1992, approximately 75,000 gallons of paint were sold or donated. The paint is collected from individuals at household hazardous waste collection days throughout California. Only paint that meets specific criteria aimed at reducing contaminants is processed to make reclaimed paint. The reclaimed paint is sold for $3.50 to $5.50 a gallon, whereas prices range from $7 to $20 a gallon for comparable new paint from manufacturers. Local California governments, community groups, volunteer organizations, and individuals can use reclaimed paint for anti-graffiti programs, the exterior of buildings, and fences. The sale of the reclaimed paint covers the costs of processing it.[26]

❖ Public Education

Governments can design their own public education materials about toxics in consumer products and ways to reduce their use and disposal, as New York City and Seattle, for example, have done. They can also make use of educational materials designed by other groups (such as the New Jersey Environmental Federation's Home SAFE Home project or Environmental Hazards Management Institute's toxics wheel).

New York and Seattle

■
Washington Toxics Coalition
Contact
Philip Dickey
Address
4516 University Way NE, Seattle, WA 98105
Phone
206-632-1545
■

Several municipalities have prepared or funded materials on household hazardous products and safer alternatives. Among these is the New York City Department of Sanitation booklet on waste reduction (described in Chapter 10) which includes such information. Another is an extensive report on metals and other toxics, mainly in household cleaning products, funded by the Municipality of Metropolitan Seattle (Metro), the city's wastewater utility, and prepared by the Washington Toxics Coalition. The 1990 report, entitled *A Database of Safer Substitutes for Hazardous Household Products*, includes detailed results of hands-on tests of the effectiveness of safer alternatives.

Home SAFE Home Program

Home SAFE Home is a consumer education program of the New Jersey Environmental Federation (a chapter of Clean Water Fund) and the Clean Water Fund nationwide. It is designed to assist consumers in making environmentally sound decisions when buying, using, and disposing of everyday products for the home and garden.

■
New Jersey Environ-
mental Federation
Address
46 Bayard St., New
Brunswick, NJ 08901
Phone
201-846-4224
■

■
Environmental
Hazards Management
Institute
Contact
Celina Y.
Grandchamp,
Educational Products
Group
Address
10 Newmarket Rd.,
P.O. Box 932,
Durham, NH 03824
Phone
603-868-1496
■

The program has created a portable "Environmental House Tour" display, an exhibit that communities can rent for a fee. It has published a quick reference chart of household alternatives to toxic products and a handbook discussing environmental actions individuals can take at home and in the community. The program's staff also work with community groups and promote "Environmental Shoppers Campaigns" to urge retail markets to provide safer, more environmentally benign products.

Household Hazardous Waste Wheel

The Environmental Hazards Management Institute (EHMI), a nonprofit environmental consulting organization, produces a variety of informational materials that governments and businesses can purchase and distribute. EHMI's Household Hazardous Waste Wheel, for example, lists chemical products, hazardous ingredients, less toxic alternatives, hazardous properties, and waste management requirements for 36 products in the general categories of auto products, household products, paints, and pesticides. Organizations that purchase the wheels can arrange to have their name and/or logo printed on them; the state of New Jersey's Hazardous Waste Facilities Siting Commission is one group that has done this.

Other Public Education Ideas

Municipalities can also adapt many of the public education strategies aimed at reducing the amount of municipal solid waste to a program to reduce the toxic constituents in solid waste. For example, they can:

- Produce and distribute educational materials for consumers (brochures, pamphlets, posters, flyers, etc.) describing strategies for reducing toxics in the waste stream.
- Publish and publicize a list of books and other resources on alternatives to products containing toxic chemicals.
- Encourage local retailers to provide information to customers about alternatives to products containing toxic chemicals.

❖ Economic Incentives and Disincentives

■
Arizona Department
of Environmental
Quality
Contact
Dennis Clayton
Address
3033 W. Central Ave.,
Phoenix, AZ 85012
Phone
602-207-4213
■

Taxes on Hazardous Substances: Arizona's Repealed "Drano Bill"

Arizona passed a bill, known as "the Drano bill," that required retailers of household products to pay an 8.5 percent tax on all products containing hazardous chemicals purchased from a manufacturer or wholesaler. However, it was repealed before it ever went into effect. The main reason for the repeal was the difficulty in determining which of hundreds of hazardous substances were in which products, without testing individual products.

❖ Regulatory Measures

While regulatory actions are often most effectively taken at the state, regional, or federal levels, municipal governments can institute some regulatory requirements: disposal bans, for example. Additionally, municipalities that recognize a need for federal or state regulations to help them accomplish their source reduction goals can press for this action. The bans and labeling initiatives discussed here provide examples of regulatory actions at a variety of levels that have targeted toxic chemicals in products.

Bans and Restrictions on Chemical Use in Products

■
CONEG Source
Reduction Task Force
Contact
Chip Foley,
Project Director
Address
Hall of States,
400 North Capitol St.,
Suite 382,
Washington, DC
20001
Phone
202-624-8450
■

CONEG Legislation A good example of legislation targeted at source reduction of toxics in the municipal waste stream is the model legislation proposed by the Coalition of Northeastern Governors (CONEG) in 1989. As discussed in Chapter 12, this legislation calls for reducing the use of four heavy metals — lead, cadmium, mercury, and hexavalent chromium — in packaging. As of mid-1992, the CONEG legislation, or legislation similar to it, had been adopted by 11 states, including seven of the nine CONEG states (Connecticut, Maine, New Hampshire, New Jersey, New York, Rhode Island, and Vermont) and four states outside the CONEG region (Iowa, Minnesota, Wisconsin, and Washington). It is pending in the two other CONEG states (Massachusetts and Pennsylvania), and many other states have it under consideration.

Environmental groups have suggested broadening this legislation by adding more metals and other substances (nickel, manganese, arsenic, titanium, copper, beryllium, cobalt, silver, gold, radioactive elements, iron, chlorine, sulfur, nitrogen, and any other significant pollutant precursors), extending similar legislation to products as well as packaging, and accelerating the deadline for phasing out metals.[27]

As also noted in Chapter 12, the proposed 1992 amendments to the federal Resource Conservation and Recovery Act include a provision, based on the CONEG model, to reduce the four heavy metals in packaging.

Ban on Batteries Containing Mercury Six states (California, Connecticut, Minnesota, New York, Oregon, and Vermont) have passed laws requiring that all batteries be mercury-free by 1996, and at least 20 other states have similar legislation pending.[28] One example of such a law is New Jersey's, signed in early 1992. It sets standards for mercury levels in dry-cell batteries and prohibits the sale and use of batteries that do not meet these standards. Household batteries must have "miniscule" mercury content to meet the New Jersey criteria. In response to such laws, battery makers have been developing batteries without mercury. However, since the laws only target mercury, the batteries may contain other toxic metals such as zinc.

■

Kemikalieinspek-
tionen
(National Chemicals
Inspectorate)
Address
P.O. Box 1384,
S-171 27,
Solna, Sweden
Phone
(46)8-730-5700

■

Swedish Ban on Cadmium Uses In 1985, Sweden banned the use of cadmium as a pigment, stabilizer, or surface treatment, but gave companies until 1992 to find substitutes. More recently, it has drawn up a plan to reduce or ban the use of 13 chemicals, including lead, mercury, and cadmium, by the year 2000.

Disposal Bans

Many states have banned the disposal of lead-acid batteries, used oil, and other products, as discussed in Chapter 12, under "Bans." Disposal bans may cause users of products to switch to less toxic alternatives or they may, alternatively, cause people to dispose of the products outside of the municipal collection and disposal system. Nonetheless, the more difficult it is to dispose of certain products, the more pressure is put on their manufacturers to reduce the toxic components.

As discussed earlier in this chapter, the regulations banning the disposal of batteries and other wastes containing mercury at Broward County's waste disposal facilities led directly to the substitution of zinc-air for mercury batteries at area hospitals. However, the pressure of a disposal ban is much less direct than bans on the sale of a product or on the use, by manufacturers, of a toxic chemical as a constituent of products.

Labeling

■

Department of
Natural Resources,
Waste Management
Assistance Division
Contact
Thomas J. Blewett,
Chief, Waste Reduc-
tion Bureau
Address
Wallace State Office
Building
Des Moines, IA
50319
Phone
515-281-8489

■

Iowa's Hazardous Product Labeling Law Iowa passed a law in 1987 requiring its Department of Natural Resources (DNR) to set up a statewide household hazardous products shelf-labeling program. The legislation covers 15 categories of product types, such as motor oils, household polishes and solvents, and petroleum-based fertilizers, and exempts eight categories (e.g., chlorine bleach, laundry detergents, cosmetics, and personal-care products). The law also requires DNR to develop a more specific list of the 50 most hazardous substances.

Although enforcement is available through the courts, DNR's primary role is to educate retailers and the public. Its Waste Management Assistance Division has the equivalent of three full-time employees working on education, setting up regional collection centers, and helping to arrange toxic cleanup day events. DNR has had since 1988 a hazardous-products hotline that permits consumers to call for information and guidance on hazardous household products. Retailers must pay a $25 annual permit fee, which helps cover program administration, toxic cleanup day activities, and educational materials that are to be displayed along with products that contain hazardous chemicals.

❖ Examples of Product Reformulation or Redesign by Manufacturers

The source reduction strategies discussed above can stimulate manufacturers to reformulate or redesign their products to contain less toxic or nontoxic constituents. The following are some examples of product reformulation or redesign that have already occurred.

Polaroid's Mercury-Free Battery for Film Cassettes The Polaroid Corporation's cameras use film cassettes containing button batteries that power the camera's electronics, optics, and film transport. Button batteries in electronic products are a major source of mercury in the waste stream. When an empty film cassette is discarded, the battery is discarded along with it. Until 1988, the Polaroid batteries, which were of the carbon-zinc type, contained both mercury and cadmium.

In 1986, the Swiss government, because of concerns about incinerator emissions, issued regulations setting limits on allowable concentrations of metals in various household batteries and requiring labels on batteries exceeding the limits to warn consumers to dispose of them separately. While Switzerland accounted for only 1 percent of Polaroid's market, its regulations were typical of guidelines being developed in other countries. According to Polaroid, as noted in a congressional Office of Technology Assessment report, the company had already decided to reduce mercury in its products prior to the Swiss regulations.[29]

In 1989, Polaroid began making a new battery that contains no mercury and reduced amounts of cadmium. The conversion to the new batteries took about two years, following several years of research.

Paint Strippers without Methylene Chloride or Other Toxic Solvents Methylene chloride is a toxic solvent traditionally used in paint strippers. Several companies, including 3M and Dumond Chemicals, Inc., have developed paint strippers that contain no methylene chloride or other toxic solvents (such as methanol, toluene, or acetone) in response to public and government concern about its health hazards, requirements that product labels list it as a carcinogen, and industry fears of a ban or restriction on its use in products.[30] While these products are more expensive and slower to use than the solvent-based products, they are safer for human health and the environment.

Product Reformulations under California's Proposition 65 Several product reformulations have occurred as a result of California's Proposition 65, a 1986 law that requires manufacturers to provide warnings on product labels if the products may expose people to chemicals (in amounts above specified risk levels) that can cause cancer or birth defects.[31] Several companies decided to reformulate products rather than provide warning labels. These companies include Gillette, which reformulated its Liquid Paper typewriter correction fluid (which contained trichloroethylene); Dow Chemical, which reformulated its K2r Spot-lifter (which contained perchloroethylene); and Pet Inc. (part of Whitman Corporation), which accelerated the elimination of lead soldering in its food cans.

Notes

1. Worldwatch Institute (Sandra Postel), *Worldwatch Paper #79, Defusing the Toxics Threat: Controlling Pesticides and Industrial Waste,* Washington, DC, 1987.

2. US International Trade Commission, *Synthetic Organic Chemicals: US Production and Sales,* 1989.

3. US Environmental Protection Agency, *EPA's Existing Chemicals Program: An Overview,* 1991.

4. US Environmental Protection Agency, Office of Toxic Substances, TSCA Hotline.

5. National Research Council, *Toxicity Testing: Strategies to Determine Needs and Priorities,* 1984.

6. University of Arizona, Environmental Systems Monitoring Laboratory, *Characterization of Household Hazardous Waste from Marin County, California, and New Orleans, Louisiana,* 1987.

7. Office of Technology Assessment, *Facing America's Trash: What Next for Municipal Solid Waste,* 1989.

8. *Ibid.*

9. *Ibid.*

10. *Ibid.*

11. US Environmental Protection Agency, *Report to Congress: Solid Waste Disposal in the United States,* Vol. II, 1988.

12. Office of Technology Assessment, *op. cit.*

13. *Ibid.*

14. *Ibid.*

15. *Ibid.*

16. New York State Legislative Commission on Solid Waste Management, *Household Hazardous Waste: An Overview,* 1988.

17. INFORM (Nancy Lilienthal, Michèle Ascione, and Adam Flint), *Tackling Toxics in Everyday Products: A Directory of Organizations,* New York, 1992. See also list of sources accompanying Tables 13-1 and 13-2 on pp. 127-128.

18. National Bureau of Standards (Churney, Ledrod, Bruce and Domalski), *The Chlorine Content of Municipal Solid Waste from Baltimore County, MD and Brooklyn, NY,* Report NBSIR 85-3213, 1985.

19. University of Arizona, Environmental Systems Monitoring Laboratory, *Characterization of Household Hazardous Waste from Marin County, California, and New Orleans, Louisiana,* 1987.

20. US Environmental Protection Agency, *A Survey of Household Hazardous Wastes and Related Collection Programs,* 1986.

21. National Solid Waste Management Association, *1987 Tipping Fee Survey,* Technical Bulletin 88-1, 1987.

22. Franklin Associates, Ltd., *Characterization of Products Containing Lead and Cadmium in Municipal Solid Waste in the United States, 1970-2000,* 1989.

23. John L. Price, "Managing Mercury Battery Wastes Through Source Substitution," *MSW Management,* January - February 1992.

24. Sally Johnson, "Recycling Is Solution for Old Paint Leftovers," *The New York Times,* March 7, 1991.

25. Personal communication, Bill McQuiggan, Owner, True Colors Home Decorating, Montpelier, Vermont, to Caroline Gelb, INFORM, September 30, 1991.

26. Personal communication, Tobi Romero, Green Alternatives, Inc., Santa Clara, California, to Caroline Gelb, INFORM, October 3, 1991.

27. Manhattan Citizens' Solid Waste Advisory Board, *List of Reduction Initiatives for Modelling in the Department of Sanitation Waste Plan,* May 1991.

28. *The Green Consumer Letter,* March 1992.

29. Office of Technology Assessment, *Facing America's Trash: What Next for Municipal Solid Waste,* 1989, based on OTA communication with H. Fatkin, Health, Safety & Environmental Affairs, Polaroid Corp.

30. *Consumer Reports,* May 1991.

31. Randolph Smith, "California Spurs Reformulated Products," *The Wall Street Journal,* November 1, 1990, p. B1.

PART III SOURCE REDUCTION PLANNING CHECKLIST

This checklist is a blueprint for action for solid waste planners who seek to reduce the amount of waste their communities generate. It details the steps state and local governments can take to create an effective source reduction program: from developing an administrative and budgetary infrastructure, to designing and implementing specific strategies, to evaluating programs and measuring success.

The checklist is designed as an organizational and planning tool that identifies basic strategies for source reduction: as such, it includes many examples of specific source reduction actions, but by no means all of the hundreds of possibilities. Thus, solid waste officials can use the checklist as an outline for developing their own source reduction plans, and as a stimulus for thinking about additional program initiatives that can meet the needs of their own communities.

The first section of the checklist, "Planning and Administration," includes the essentials of creating a source reduction program. The second section, "Basic Source Reduction Strategies," identifies dozens of specific initiatives that virtually all waste-generating sectors (governments, businesses, institutions, residents) can take to reduce waste: by changing the products they buy (procurement) and by changing how they use these products (operations). The remainder of the checklist itemizes additional strategies solid waste planners can use: stimulating source reduction through technical assistance, pilot programs, information clearinghouses, grants, and other activities; educating consumers and students; creating economic incentives and disincentives to promote source reduction; and using regulatory measures. A final section identifies initiatives for reducing toxic chemicals in the waste stream.

While the checklist is intended primarily for government planners, it can also be used by citizen and environmental groups seeking to promote government source reduction action, and by businesses and institutions interested in developing source reduction programs in their own organizations.

I. Planning and Administration

Before source reduction planners can start developing specific source reduction initiatives for their communities, they need a clear definition of source reduction, an explicitly stated source reduction policy, clearly defined goals, meaningful measurement strategies, an effective administrative structure, and an adequate budget.

✔ Policy

❑ Designate source reduction as the top priority in solid waste policy in accordance with the EPA-endorsed hierarchy of solid waste policy options.

✔ Definition

❑ Define waste management terms so that the distinctions among the terms "source reduction," "recycling," "waste reduction," and "disposal" are clear.
  ❑ Source reduction (or waste prevention): reduction of the amount and/or toxicity of waste actually generated
  ❑ Recycling: remanufacturing of materials collected from the waste stream to make new products
  ❑ Waste reduction: reduction in the amount of waste requiring disposal; includes both source reduction and recycling
  ❑ Disposal: managing waste through methods such as incineration and landfilling

✔ Waste Audits

Audit information makes it possible to target specific materials or toxic substances for reduction because they are:
  ❑ Major contributors to the waste stream
  ❑ Difficult to recycle
  ❑ Easy to reduce, or
  ❑ Major contributors to pollution during the disposal process

Audit information also makes it possible to set realistic goals for source reduction. Governments can:

❑ Conduct a waste audit for planning and evaluation purposes. The audit should determine:
  ❑ The amount of waste
  ❑ Waste sources
  ❑ Waste composition

✔ Goals and Measurement

❑ Establish an overall source reduction goal that is separate from the recycling goal, with specification of the:
  ❑ Baseline year
  ❑ Target year
  ❑ Type of reduction to be measured: from current total waste generation levels, from current per capita generation levels, or from projected increase

❑ Determine separate goals desired for:
  ❑ Generating sectors: residential, commercial, institutional
  ❑ Materials: paper, glass, plastics, organics, etc.
  ❑ Products: styrofoam, tires, cardboard, newspapers, etc.

❑ Select unit of measurement:
  ❑ Weight
  ❑ Volume
  ❑ Weight and volume (preferable, if possible)

❑ Select measurement methodology:
  ❑ Waste audits
  ❑ Sampling
  ❑ Surveys
  ❑ Purchases (tracking sales)

❑ Improve data collection and reporting so that measurement of goals set is possible, by:
  ❑ Disaggregating data by generating sectors: residential, commercial, institutional
  ❑ Disaggregating data by basic material: yard waste, bottle bill material, bulky items, etc.
  ❑ Quantifying waste exports and/or imports, and waste collected and handled by private carters
  ❑ Performing waste audits
  ❑ Providing scales at disposal facilities to weigh materials collected

✔ Administration

❑ Create a source reduction staff that is, preferably, separate and independent of recycling and disposal functions. Staff is needed to work on planning, program development and evaluation, legislation, data collection, waste audits, technical assistance, and education and outreach.

✔ Budget

❑ Designate a budget for source reduction that is distinct from the recycling budget.

❑ Allocate funds to design, implement, publicize, and evaluate source reduction programs.

II. Basic Source Reduction Strategies

Once a community has established its source reduction goals, developed a method of tracking progress, and created a source reduction office with staff and budget, the next step is to develop and implement specific strategies aimed at reducing the amount and toxicity of waste. State and local governments can tailor two basic approaches to virtually all waste-generating sectors (government, businesses, institutions, residences); these approaches entail changing procurement policies and changing daily habits or operations.

Government. One out of every six workers in the United States is employed by federal, state, or local government, a total of 19 million people. And, each year, federal, state, and local governments spend almost $1 trillion — approximately 20 percent of the gross national product. Reducing waste generated by this sector could have a great impact on the municipal solid waste stream. While exact figures are not available, government generates approximately 12 million tons of waste each year, since the nonresidential sector generates about 72 million tons of waste annually.

Businesses. With the commercial sector generating about 40 percent of this country's waste, businesses have many opportunities to reduce their wastes. Government can adapt business initiatives for government offices and operations, can encourage businesses to adopt procurement and other strategies that have been effective in the government sector, and can publicize business source reduction accomplishments to stimulate source reduction initiatives in other businesses and other waste-generating sectors.

Institutions. Institutions, such as correctional, health care, educational, and cultural facilities, also generate substantial amounts of waste. For New York City, this waste has been estimated at over 8 percent of the total waste stream. State and local governments can implement programs in the institutions they operate and encourage privately run institutions to replicate these efforts.

Residents. The residential sector generates about 60 percent of all municipal solid waste in the United States — some 108 million tons. As consumers, residents can change what they buy and what they do in order to generate less waste; hence, many of these procurement and operational strategies can be adapted for residential waste generators as well. Other source reduction strategies aimed at consumers and students are outlined in Section IV.

✔ Procurement

Incorporating source reduction objectives into procurement policy guidelines can decrease the amount of waste generated, encourage manufacturers to develop and sell products that create less waste, and often save money. Some general categories of procurement changes that promote source reduction, as well as some of the many specific examples of changes that can be made, are listed here.

❑ Buy reusable items instead of single-use items, such as:
  ❑ Office supplies
    ▪ Refillable pens and pencils
    ▪ Refillable laser toner printer cartridges
    ▪ Multistrike typewriter ribbons
    ▪ Envelopes with metal clasps
    ▪ Erasable wall calendars
    ▪ Washable towels
  ❑ Food service
    ▪ Flatware, mugs, dishes, trays
    ▪ Coffee filters
    ▪ Salt and pepper shakers
    ▪ Bulk containers for sugar, sugar substitutes, mustard, ketchup, etc. (to avoid individually wrapped servings)
  ❑ Health care facilities
    ▪ Patient gowns
    ▪ Bed pads
    ▪ Pillows

▪ Cloth diapers
▪ Medical instruments, where it is deemed safe to do so
▪ Materials for in-house day care programs
▪ Trash containers

❑ Use fountain soda dispensers, milk in steel containers, and water coolers to decrease use of disposable beverage containers, and purchase other beverages, when possible, in refillable bottles.

❑ Buy items in bulk, such as cleaning products, food products, and office supplies.

❑ Buy concentrated cleaning supplies (e.g., detergents) and food items (e.g., juices, soups).

❑ Increase use of durable and repairable materials by:
  ❑ Buying or leasing durable and repairable equipment, such as photocopiers, fax machines, computers, typewriters, and coffee-makers
  ❑ Negotiating for longer and more comprehensive warranties and service contracts when purchasing durable products
  ❑ Buying sturdy office supplies, such as staplers, scissors, file holders, and book ends

❑ Use procurement clout to:
  ❑ Encourage manufacturers to provide products that promote source reduction, such as photocopy machines that are programmed to automatically make double-sided copies.
  ❑ Require suppliers to use minimal packaging, reusable shipping containers, and reusable packaging material (such as scrap paper instead of new packing peanuts, and blankets for furniture), and/or take back the containers or packages.

❏ Adopt procurement strategies that promote paper waste reduction, such as buying:
  ❏ Laser printers that can make double-sided copies
  ❏ Computer programs that permit faxing directly from a computer to avoid print-outs
  ❏ Fax machines that use plain paper (since most faxed documents are photocopied and the fax is then discarded)
  ❏ Electronic mail systems
  ❏ Narrow-lined notepads
  ❏ Two-way envelopes for mailings that require returns or responses, such as bills
  ❏ Reusable envelopes for interdepartmental mailings
  ❏ Fewer copies of publications and telephone books

❏ Give a price preference for reusable, refillable, durable, and repairable equipment, and for less toxic products.

❏ Publicize source reduction procurement practices through trade purchasing organizations to encourage other government agencies, businesses, and institutions to adopt source reduction measures.

❏ Purchase items that have fewer toxic constituents, such as:
  ❏ Unbleached paper products
  ❏ Low-toxic correction fluid or correction tape
  ❏ Products with less toxic inks (soy-based) and dyes
  ❏ Reclaimed (remixed) latex (water-based) paints
  ❏ Cleaning products with nontoxic constituents
  ❏ Equipment that does not require batteries (such as solar-powered calculators and manual pencil sharpeners), or that uses rechargeable batteries

❏ Solid wood instead of particle board, and carpet tacks instead of floor-covering adhesives with toxic constituents, when redesigning or moving offices

(For further information on reducing toxic chemicals in the municipal solid waste stream, see Section VII.)

✔ Operations

Source reduction can also be achieved by encouraging changes in how goods are used and disposed of. The first priority is avoiding the use of some materials in the first place (for example, circulating a memo instead of distributing individual copies to five people saves four copies). In addition, materials can be reused, eliminating the need to buy, use, and dispose of additional items. The use of materials can also be "intensified," to keep them in working condition longer and get the maximum use out of a product. For instance, repairing broken furniture, office equipment, and home appliances can save money and keep items out of the waste stream. Some general categories of operational changes that support source reduction, as well as some specific examples of changes that can be made, are listed here.

❏ Reduce office paper waste by:
  ❏ Preparing documents double-sided and single-spaced
  ❏ Changing margins to avoid pages with little text
  ❏ Editing and proofreading on the computer before printing
  ❏ Loading laser printer paper trays with paper already used on one side for drafts and internal memos
  ❏ Circulating/sharing documents, publications, and telephone books
  ❏ Posting office announcements on bulletin boards

- ❑ Setting up central filing systems
- ❑ Storing files on computer disks
- ❑ Using small pieces of paper for short memos
- ❑ Using paper already used on one side for drafts and notes
- ❑ Eliminating fax cover sheets
- ❑ Increasing use of electronic mail

❑ Reduce direct mail in the waste stream by:
- ❑ Targeting specific audiences for direct mail as precisely as possible to reduce the amount sent
- ❑ Eliminating duplication by frequently updating mailing lists
- ❑ Removing name from unwanted or duplicate mailing lists

❑ Reduce packaging waste by:
- ❑ Shipping and mailing materials in minimal packaging, reusable packages, or no packaging at all, when feasible

❑ Reduce organics in the waste stream by:
- ❑ Composting yard waste on site
- ❑ Leaving grass clippings on lawns, or mulching grass and leaves
- ❑ Reusing wood from trees and bushes as mulching chips
- ❑ Donating wood to be used as firewood

❑ Extend equipment life through:
- ❑ Proper maintenance
- ❑ Repairs instead of replacements

❑ Promote employee attention to waste reduction through:
- ❑ Education (posters, newsletter articles, etc.)
- ❑ Incentives (such as price discounts for using reusable tableware in cafeterias)
- ❑ Swaps of items from home that are no longer needed

❑ Donate items that are no longer needed (equipment, furniture, food, wood, etc.) to organizations that can use or sell them.

❑ Conduct waste audits and materials assessments in non-administrative sectors of government agencies and businesses to find additional opportunities for source reduction.

❑ Reduce toxicity of waste by:
- ❑ Remixing and reusing latex paint
- ❑ Using nontoxic materials in place of toxic chemical-containing commercial cleaners and pesticides (such as vinegar and water for cleaning glass, baking soda and water for polishing silver and stainless steel, and liquid pure soap and water for controlling garden pests)

❑ Reduce the amount of tires in the waste stream, by:
- ❑ Keeping tires filled to proper air pressure for maximum life
- ❑ Having truck and passenger vehicle tires retreaded
- ❑ Using high-mileage tires

❑ Use re-refined lubricating oil

III. Government Programs to Stimulate Source Reduction

✔ Technical Assistance

Government technical assistance programs can help businesses and institutions recognize opportunities for and implement source reduction programs. Specifically, they can help businesses conduct waste audits and materials assessments, recognize opportunities for reducing waste, and implement source reduction programs. Government programs can be run by government employees or through volunteers from within the private sector. As components of a technical assistance program, governments can:

❏ Provide businesses with information about how to conduct waste audits and materials assessments to identify opportunities for reducing waste. To conduct such an audit, businesses gather data about:

  ❏ Materials that are purchased by the business
  ❏ Packaging in which purchased materials are shipped
  ❏ Materials brought into the office by employees
  ❏ Information received in the mail (e.g., letters, bills, documents, direct mail, envelopes, and packaging)
  ❏ Items shipped from the office, including products and their packaging, reports, direct mail, etc.
  ❏ Materials kept in the office (such as in the supply room and filing cabinets) that may be discarded later
  ❏ Materials that immediately become waste (such as food and its packaging, and envelopes)
  ❏ Materials separated for recycling

❏ Assist businesses in setting up source reduction programs by:

  ❏ Surveying businesses to inquire about their waste management practices and to find out whether they would like to participate in a pilot waste prevention program.
  ❏ Providing businesses with information and other assistance throughout the multistep process of creating and implementing a source reduction program. Steps in this process include:

    ▪ Getting upper management support and distributing a policy statement to all employees

    ▪ Creating a source reduction task force with representatives from all different departments (i.e., administration, janitorial, purchasing, and professional staff)
    ▪ Gathering basic waste generation and material usage information from each department and reporting current purchases (amounts and costs) to develop baseline data
    ▪ Setting goals (company-wide, departmental, or by material)
    ▪ Setting an agenda by:

      ▪ Putting out suggestion boxes for employees
      ▪ Gathering ideas from task force members
      ▪ Prioritizing strategies based on those that are easiest to implement and those that will reduce the greatest amount of waste and save the most money
      ▪ Discussing obstacles and ways to overcome them

    ▪ Implementing programs through task force members
    ▪ Expanding and evaluating the program by discussing problems and developing solutions; continuing to test new strategies; and measuring, evaluating, and documenting cost and disposal savings

❏ Encourage participating businesses to publicize their source reduction activities in trade magazines to stimulate other businesses to adopt source reduction practices on their own.

❏ Publicize information generated by participating businesses to help other businesses see the financial savings from adopting source reduction practices.

❑ Evaluate business efforts to identify where improvement can be made to further reduce waste.

✔ Backyard Composting and Leave-on-Lawn Programs

Government assistance can be used to reduce the amount of yard waste in the waste stream by promoting backyard composting and leave-on-lawn programs. Yard waste, which consists of leaves, branches, and grass clippings, is the second largest material category in the US waste stream — 17.6 percent by weight in 1988 (after paper, at 40 percent). Food waste comprises 7.4 percent of the waste stream. These materials can be reduced at the source by backyard composting, leaving grass clippings on the lawn, and mulching grass and leaves. In addition to reducing waste, keeping these materials out of the waste stream could reduce pollution from disposal systems since yard and food wastes may create nitrogen oxides when burned, and methane gas, leachate, and settling problems when landfilled. To reduce the amount of these organics in the waste stream, governments can:

❑ Educate residents and businesses about the benefits of leaving grass clippings on lawns or mulching grass and leaves.

❑ Teach residents and businesses how to compost, using brochures, conferences, demonstrations sites, and/or in-home training.

❑ Distribute free composting bins to residents.

❑ Encourage development of urban community gardens that use composted materials as fertilizer.

✔ Reuse Programs

Governments can sponsor, promote, and publicize reuse programs that distribute used goods. These programs keep materials that would otherwise be discarded out of the waste stream and make items available at lower costs.

❑ Examples of reuse programs are:
  ❑ Yard sales
  ❑ Food distributions to drug treatment centers, day care centers, homeless shelters, and senior citizen homes
  ❑ Donations of furniture and other materials to arts and other nonprofit organizations, schools, hospitals, etc.
  ❑ Thrift shops that sell used home furnishings, supplies, and clothing

✔ Pilot Programs

Government-sponsored pilot programs can provide an opportunity to test and demonstrate source reduction strategies. To implement pilot programs, governments first need to select businesses or residential areas that are representative of the community so successful source reduction measures can be easily replicated elsewhere. They then can:

❑ Gather data on present waste generation habits and amounts.

❑ Target specific large waste contributors.

❑ Measure and document cost savings and reductions in waste to facilitate expanding and improving the pilot program(s) into a comprehensive municipal program.

❑ Publicize findings to encourage others to adopt measures on their own or to improve existing programs.

Governments can establish clearinghouses that make information about source reduction facts and issues available to community activists, businesses, residents, and government officials. Having collected the information, government-sponsored clearinghouses can:

❏ Publicize the availability of the information.

❏ Distribute informational materials (fact sheets, pamphlets, reports, etc.) that outline for businesses and institutions how to include source reduction criteria in their procurement policies and programs.

❏ Distribute information about the benefits of and strategies for source reduction at local environmental and business conferences and events.

❏ Publicize good government, business, institutional, or residential source reduction programs (and their waste and cost savings) on local television and radio stations and in local newspapers.

❏ Make waste education newsletters available to businesses and institutions.

Contests and awards can encourage businesses, institutions, and manufacturers to adopt source reduction practices and to increase consumer awareness about source reduction opportunities. Governments can:

❏ Give awards recognizing businesses, individuals, or organizations that:
   ❏ Achieve significant reductions in the amount of waste they generate in their manufacturing and administrative processes.
   ❏ Develop product designs that accomplish source reduction.
   ❏ Change their packaging in a way that reduces waste.
   ❏ Increase their purchase and use of durables.

❏ Hold contests to generate citizen involvement in developing source reduction campaign slogans.

State and local governments can give grants to stimulate businesses and institutions to develop innovative strategies for reducing waste on their own. They can:

❏ Designate grant money for source reduction research and projects.

❏ Publicize findings of grant projects to encourage other organizations to adopt source reduction practices on their own.

IV. Educating Consumers and Students

The residential sector generated approximately 108 million tons of waste in 1988; this is expected to increase to 150 million tons by the year 2000 if source reduction measures are not practiced. The approximately 60 million students enrolled in kindergarten through college, and school faculty and staff, are estimated to generate some 5 to 7 million tons of waste annually. Local governments can publish and distribute a variety of educational materials that give individual consumers specific recommendations on how to reduce waste and can encourage schools to develop curricula and adopt practices that promote source reduction.

Governments can:

❑ Design and distribute brochures, pamphlets, and posters that give consumers specific recommendations on how to adopt less wasteful practices and make less wasteful purchasing decisions, such as those listed under "Basic Source Reduction Strategies." Examples of recommended practices can include:

❑ Bringing reusable shopping bags to stores

❑ Returning plastic bags and hangers to dry cleaners

❑ Donating items no longer needed to charitable and social service organizations

❑ Reusing containers and other materials

❑ Maintaining items for longest life (e.g., keeping proper tire pressure, cleaning air conditioners)

❑ Buying perishable items only when needed and checking expiration dates when purchasing such products to avoid the need to dispose of spoiled or unused items

❑ Distribute directories listing repair shops, thrift shops, and charity groups that accept donations and sell second-hand goods.

✔ Educating Students

Schools are an effective place to begin source reduction programs because reaching children at an early age is a good way to encourage good environmental habits. Further, since school waste represents 2.6 to 4 percent of US waste, reducing it can have a significant impact on the overall waste stream. Schools can:

❑ Develop waste education curricula that focus on reduction and reuse, including such activities as:

❑ Developing worksheets and videos to help students understand waste problems and ways to solve them through source reduction

❑ Conducting mini-composting and decomposition projects

❑ Collecting daily trash to visualize the amount generated and to develop opportunities for elimination, reduction, and reuse

❑ Developing source reduction promotional campaigns

❑ Using items that would otherwise be discarded in arts and crafts projects

❑ Establish an extracurricular club to focus on identifying and implementing source reduction opportunities within the school.

❑ Distribute source reduction fact sheets to students (and parents) giving them ideas for reducing waste.

❑ Increase use of bulletin boards and overheads to reduce handouts.

❑ Place swap boxes in schools or designate swap days for students to bring materials and equipment from home that would otherwise be discarded.

❑ Collect unused school materials at the end of the year to donate to other students at the beginning of the next year.

V. Economic Incentives and Disincentives

✔ Variable Waste Disposal Fees

Variable waste disposal fee systems that charge residents based on the amount of waste set out for disposal create an incentive to reduce and recycle since the more garbage a household generates, the more it must pay for its disposal. To reduce waste through variable waste disposal fee systems, governments need to:

❏ Include provisions that enable residents to save more by reducing waste than by recycling.

❏ Include specific provisions for multi-tenant buildings and low-income households.

❏ Educate citizens about the importance of reducing waste in order to maximize reduction rates.

❏ Establish an enforcement mechanism for illegal dumping (e.g., teams of inspectors, reporting hotlines, etc.).

✔ Taxes

Taxes on products and packages can be designed to encourage consumers to purchase products that generate less waste, help pay for the disposal of products, encourage manufacturers to adopt source reduction measures, and raise funds for other solid waste activities. Governments can place taxes on:

❏ Waste-producing products, such as non-refillable containers

❏ Products that use unnecessary packaging (whether it is virgin or recycled material), such as packages that exceed a specified product-to-package ratio or that have multiple layers of packaging

✔ Deposits/Refunds

Deposit/refund laws place fees on products when they are purchased and give the deposit back to the consumer upon return of the package or product. Governments can:

❏ Encourage local manufacturers to develop deposit/refund systems for reusable items

such as refillable beverage bottles, food containers, paint buckets, and shampoo bottles.

❏ Support state and national efforts to develop deposit/refund systems on a larger scale.

❏ Encourage state and federal governments to set refill rates for beverage bottles.

❏ Design bottle bills that encourage refilling over recycling and recycling over disposal.

❏ Educate consumers to buy and return products that can be returned for refund and then reused.

✔ Tax Credits

Governments can give tax credits to:

❏ Manufacturers that reduce packaging, change to reuse/refill systems, increase product durability, or eliminate or reduce a toxic component of a product or package

❏ Businesses that purchase equipment or materials that will help reduce waste, such as double-sided photocopy machines, equipment for a cloth diaper cleaning service, or electronic mail systems

✔ Financial Bonuses

To encourage source reduction at the disposal level, governments can:

❏ Offer financial incentives to carters/haulers or to communities to reduce the amount of waste they bring to incinerators or landfills. (Such incentives often encourage recycling and source reduction equally, but can be designed to give priority to source reduction.)

VI. Regulatory Measures

✔ Required Source Reduction Plans

Writing source reduction plans can help businesses identify waste reduction opportunities and possible financial savings. Local governments can require businesses and institutions to:

❏ Conduct waste audits and materials assessments, and follow other recommendations listed under "Technical Assistance" in Section III.

❏ Write source reduction plans.

❏ Submit regular updates on progress.

✔ Labeling Programs

Labeling programs can help consumers determine which products and packages create the least waste. To implement such programs, governments need to:

❏ Determine if the labeling program will involve labels for products or shelves.

❏ Develop specific, consistent criteria for labeling standards.

❏ Encourage retail participation: for example, by listing participants in (or violators of) the program in local newspapers.

❏ Establish an enforcement mechanism, such as using a hotline for consumers to report violators so that inspectors can investigate complaints.

❏ Publicize information about local programs to support development of national programs.

✔ Bans

Disposal and retail bans on specific materials or items can encourage manufacturers to develop alternative products or packages that create less waste. Governments can:

❏ Ban materials and items from disposal that either cause problems when landfilled or incinerated because they are major contributors to the size and toxicity of the waste stream, or can easily be eliminated, reduced, or reused (e.g., batteries, tires, glass, metal, major appliances, yard waste, used oil, drain cleaners, solvent, paints, disposable cups, disposable diapers, and grocery bags).

❏ Publicize bans and educate citizens about why items are banned and what they can do to keep these items out of the waste stream and out of recycling systems (e.g., backyard composting of yard waste, donations of unused paints).

❏ Educate consumers to make them aware of alternatives to banned products (e.g., homemade cleaners, reusable cups, cloth diapers, and rechargeable batteries).

✔ Packaging Initiatives

On average, each resident of the United States generates 463 pounds of packaging waste per year; packaging comprises one-third of the national waste stream by weight. While some packaging is essential for containing, protecting, transporting, and marketing products, significant reductions in waste generation and costs can be achieved by eliminating unnecessary packaging, reusing/refilling, and designing more efficient packages. To reduce packaging waste, governments can:

❑ Encourage retail stores to create incentives for customers to carry items in reusable bags, such as charging for disposable bags.

❑ Work with other state and local governments to:
  ❑ Develop packaging standards.
  ❑ Make changes in current procurement requirements.
  ❑ Encourage the nongovernment sector to adopt the same standards.

VII. Reducing Toxic Chemicals in the Waste Stream

From paints, batteries, and home pesticides to cleaning products, fingernail polish, and motor oil, there are a variety of consumer products containing synthetic and other chemicals that are known to be or may be harmful to our health or the environment. When such products are disposed of, these chemicals make their way to our landfills, garbage incinerators, home septic systems, and public sewer systems. From there, they can move into our air, land, and water.

The most important strategy for reducing toxic chemicals in the waste stream is to encourage the purchase and use of products containing fewer toxic constituents than those commonly in use. (People who already have toxic products in their homes or businesses should be encouraged to use them up or give them away, instead of throwing them out; however, the use of products containing toxic chemicals can cause other environmental problems, such as indoor air pollution and smog.) To promote reduction of toxics in the waste stream, governments can:

❑ Set source reduction goals for reducing the purchase and disposal of specific products that are known to contain particular toxic substances, including:
  ❑ Identifying chemicals and products of concern
  ❑ Identifying existing less toxic alternatives
  ❑ Setting goals by product and chemical

❑ Measure progress in reducing toxic chemicals in the waste stream. Options include:
  ❑ Monitoring the sales of targeted products in the municipality
  ❑ Monitoring activity at product exchange centers to develop information on how much of unused targeted products is being acquired and used up
  ❑ Sampling the waste stream for the targeted products

❑ Change procurement policies and operations. Some examples are:
  ❑ Requiring paper contracts to specify unbleached paper
  ❑ Purchasing less toxic cleaning supplies and less toxic pesticides
  ❑ Remixing and reusing latex paint

❑ Design and implement programs for reducing toxics for specific sectors, such as encouraging hospitals to use zinc-air batteries instead of mercury batteries in cardiac monitors and intermediate care units, and encouraging schools to use nontoxic art supplies.

❑ Publicize, promote, or sponsor reuse programs that reduce toxics in the waste stream.

❑ Design consumer education programs that include such activities as:
  ❑ Producing and/or publicizing reference charts of household alternatives to products containing toxic chemicals

❏ Distributing information on reducing toxics in the waste stream to residents and businesses, including references to handbooks that discuss environmental actions consumers can take

❏ Establish other programs to encourage manufacturers to develop products with fewer toxic constituents and to encourage people to buy or make alternatives containing fewer toxic chemicals, such as:

❏ Taxing manufacturers and/or retailers for specific products known to contain hazardous chemicals

❏ Banning specific toxic chemicals in packaging and products

❏ Banning disposal of products containing specified toxic chemicals

❏ Requiring retailers to label products containing specified toxic chemicals

Bibliography

Albany Medical Center (Claude D. Rounds). *Cure Waste*. Presented at the New York State Department of Environmental Conservation Third Annual Recycling Conference. Rochester, New York: November 13, 1991.

Allen, Frank Edward. "Environmental Terms Catch on Very Slowly." *The Wall Street Journal*. July 11, 1991.

Arthur D. Little, Inc. *A Report on Advance Disposal Fees*. Prepared for Environmental Education Associates. Cambridge, Massachusetts: April 1992.

Beer Institute. *1990 and 1982 Estimate Draugt and Container Share by State*. Washington, DC: 1991.

Cal Recovery Systems, Inc. (Marian R. Chertow). *Final Report: Waste Prevention in New York City: Analysis and Strategy*. Prepared for the New York City Department of Sanitation. January 15, 1992.

Chapin, Scott. "The Return of Refillable Bottles." *Resource Recycling*. March 1991.

Cisternas, Miriam G. and Rosemary Swanson. "Source Reduction for Municipalities: An Agenda for Action." University of California at Los Angeles Masters Project: 1991.

Clarke, Marjorie J. "The Paradox and the Promise of Source Reduction." *Solid Waste & Power*. February 1990.

Coalition for Northeast Governors.

"Collapsible Containers to Save $1 Million." *Transportation & Distribution Magazine*. March 1991.

Concern, Inc. *Waste: Choices for Communities*. Washington, DC: September 1988.

Congressional Research Service Report for Congress (James E. McCarthy). *Recycling and Reducing Waste: How the United States Compares to Other Countries.* November 8, 1991.

Connecticut Department of Adminstrative Services. *Plan to Eliminate Disposables and Single Use Products in State Government.* February 1, 1990.

Connecticut Department of Environmental Protection (Judy Belaval). *Connecticut Recycles Office Paper.* Jannuary, 1990.

Consumer Reports. May 1991.

Cornell Waste Management Institute and Tomkpins County Division of Solid Waste (Sarah Stone). *A Final Report: Tompkins County Trashtag and Recycling Study.* 1991.

D'Allessandro, Bill. "The Green Cross Certification Company (Oakland, California) Will Offer a Product Eco-Label in the US Based on the Life Cycle Analysis." *Crosslands European Environmental Bulletin.* New Hampshire: September 12, 1991.

Delaware Department of Natural Resources. *Delaware Home Audit Kit: A Guide to Help Make Your Residence an Environmentally Friendly Place.*

Environmental Data Services, Ltd. *The ENDS Report.* Surrey, England: October 1991.

Environmental Defense Fund (Richard A. Denison and John Ruston). *Recycling and Incineration Evaluating the Choices.* Washington, DC: 1990.

"Federal Watch." *Resource Recycling.* July 1991.

Fibre Box Association. *Fibre Box Handbook.* 1989 Edition.

Franklin, Associates, Ltd. *National Office Paper Recycling Project: Supply of and Recycling Demand for Office Waste Paper 1990 to 1995.* Kansas: 1991.

Franklin, Associates, Ltd. *Characterization of Products Containing Lead and Cadmium in Municipal Solid Waste in the United States, 1970-2000.* Kansas: 1989.

Freudenheim, Milt. "The Tiniest, Kindest Cut of All." *The New York Times.* July 10, 1991.

Goldberg, Dan. "The Magic of Volume Reduction." *Waste Age.* February 1990.

Goodwill Industries of America, Inc. *Facts.* Washington, DC: 1990.

Goodwill Industries of America, Inc. *1990 Annual Report.* Washington, DC: 1990.

Grant, Phyllis, Susan Laird and Doris Coppage. "Disposable Versus Reusable Pillows: A Case Study." *Hospital Material Management Quarterly.* February, 1990.

Green Consumer Letter. May 1992.

Gunderson, Kimberly. "Iowa's IWRC 'Shows Small Businesses How.'" *Pollution Prevention*. March 1991.

Hamilton County Solid Waste Management District News. Wyoming: Second Quarter 1991.

Herrmann, Gretchen Mary. "Garage Sales As Practice: Ideologies of Women, Work and Community in Daily Life" Abstract of dissertation. State University of New York at Binghamton: 1990.

Holusha, John. "Mixed Benefits From Recycling." *The New York Times*. July 26, 1991.

Illinois Department of Energy and Natural Resources (Elliott Zimmermann). *Solid Waste Management Alternatives: Review of Policy Options To Encourage Waste Reduction*. February 1988.

In Business. "A Packaging Ban." Summer 1991.

In Business. "Bottom Care." Summer 1991.

INFORM (Marjorie J. Clarke, Maarten de Kadt, Ph.D., and David Saphire). *Burning Garbage in the US: Practice vs. State of the Art*. New York: 1991.

INFORM (Maarten de Kadt, Ph.D.). *Recycling Programs in Islip, New York, and Somerset County, New Jersey*. New York: 1991.

INFORM (Robert Graff and Bette Fishbein). *Reducing Office Paper Waste*. New York: November 1991.

Institute for Local Self-Reliance (Brenda Platt, *et al.*). *Beyond 40 Percent: Record Setting Recycling and Composting Programs*. Washington, DC: 1990.

Integrated Waste Management. "National Solid Waste Management Association (NSWMA) Survey Tracks Upward Trend in Landfill Tipping Fees." New York: December 11, 1991.

Johnson, Sally. "Recycling Is Solution For Old Paint Leftovers." *The New York Times*. March 7, 1991.

"Legislative Initiatives to Reduce Municipal Solid Waste." *Resource Recycling*. October 1990.

Maine Waste Management Agency. *Maine Waste Management and Recycling Plan*. July 1990.

Manhattan Citizens' Solid Waste Advisory Board. *List of Reduction Initiatives for Modelling in the Department of Sanitation Wste Plan*. May 1991.

Martin, Amy. "Airport Recycling." *Resource Recycling*. February 1991.

Michigan Department of Natural Resources. *Protecting Michigan's Future Bond Solid Waste Alternatives Projects*. March 12, 1991.

Michigan Department of Natural Resources. *Michigan Waste Prevention Strategy*. February 14, 1991.

Michigan Department of Natural Resources. *Michigan Solid Waste Policy*. June 1988.

Miller, Michael W. "'Greens' Add to Junk Mail Mountain." *The Wall Street Journal*. May 13, 1991.

Minnesota Governor's Office. *Select Committee on Packaging and the Environment: Final Report*. December 18, 1990.

Minnesota Department of Administration, Division of Materials Management. *Resource Recovery Biannual Report: FY 1989 and 1990*. Submitted to the Legislative Commission on Waste Management. January 1991.

Minnesota Office of Waste Management. *Examples of Source (Waste) Reduction by Commercial Businesses*. March 1989.

Minnesota Office of Waste Management. *Results of Phase 1: Itasca County Solid Waste Reduction Pilot Project*. June 11, 1990.

Minnesota Office of Waste Management. *Solid Waste Reduction and Reycling Programs*. May 1990.

Minnesota Office of Waste Management. *Waste Source Reduction - A Hospital Case Study*. 1992.

Minnesota Office of Waste Management. *Waste Source Reduction - Small Business Case Study*. 1991.

Nader, Ralph, Eleanor J. Lewis and Eric Weltman. *Government Purchasing Project Testimony delivered to the Subcommittee on Oversight of Government Management of the Senate Government Affairs Committee*. November 8, 1991.

National Bureau of Standards (Churney, Ledrod, Bruce and Domalski). *The Chlorine Content of Municipal Solid Waste from Baltimore County, MD and Brooklyn, NY*. Washington, DC: 1985.

National Research Council. *Toxicity Testing: Strategies to Determine Needs and Priorities*. Washington, DC: 1984.

National Solid Waste Management Association. *Recycling in the States: 1990 Review*. Washingon, DC: 1991.

National Solid Waste Management Association. *1987 Tipping Fee Survey*, Technical Bulletin 88-1. Washington, DC: 1987.

National Wildlife Federation. *Citizens Action Guide*.

New Jersey Executive Department. *The Emergency Solid Waste Assessment Task Force: Final Report*. August 6, 1990.

New York City Department of Sanitation. *A Comprehensive Solid Waste Management Plan for New York City and Final Generic Environmental Impact Statement*. August 1992. (Also, March 1992 draft.)

New York City Department of Sanitation. *Waste Prevention Partnership Press Kit*. September 25, 1991.

New York City Department of Sanitation. *Collectors' Items*. Fall, 1991.

New York City Department of Sanitation. *New York City Recycles: Preliminary Recycling Plan, Fiscal Year 1991*.

New York State Department of Environmental Conservation. *Draft New York State Solid Waste Management Plan: 1990/91 Update*. October 1990.

New York State Legislative Commission on Solid Waste Management. *Household Hazardous Waste: An Overview*. Albany, New York: 1988.

Office of Technology Assesment. *Facing America's Trash: What Next for Municipal Solid Waste*. Washington, DC: 1989.

Olmsted County, Minnesota. *Solid Waste Management Plan Update*. December 1989.

Oregon Department of Environmental Quality. *Amended 1986 Waste Reduction Program*. March 1989.

Organisation for Economic Co-operation and Development. *OECD Environmental Data Compendium 1991*. Paris: 1991.

Pennsylvania Department of Environmental Resources. *Act 101: Waste Reduction Study*. September 1990.

Powell, Jerry. "State Recycling Coordinators: 1991 Salary Survey." *Resource Recycling*. June 1991.

Price, John L. "Managing Mercury Battery Wastes Through Source Substitution." *MSW Management*. January/February 1992.

"Programs in Action." *Resource Recycling*. July 1991, August 1991, February 1992.

Reid, Patti L. "Students Push Cafeterias To Recycle and Re-Use." *The New York Times*. September 15, 1991.

Research Innovations for The Hartman Public Relations Group (Karen A. Brattesani, Ph.D.). *1990 Waste Reduction Survey*. Prepared for Seattle Solid Waste Utility. April, 1990.

"Re-Use a Good Sign for Goodwill." *Retailing & The Environment/Discount Store News*. March 18, 1991.

"Reuse Strategy Penalizes Plastic." *Plastics News*. December 9, 1991.

Rhode Island Solid Waste Management Corporation. *Report on Waste Reduction in Rhode Island State Agencies*. August 1991.

Rhode Island Department of Environmental Management. *Ocean State Cleanup and Recycling and Guide for Preparing Commercial Solid Waste Reduction and Recycling Plans*. 1988.

Rigdon, Joan E. "For Cardboard Cameras, Sales Picture Enlarges and Seems Brighter than Ever." *The Wall Street Journal*. February 11, 1992.

Roach, Margaret. "Mow It, but 'Don't Bag It.'" *New York Newsday*. March 21, 1991.

Rukikoff & Rohde, Inc. *Draft Generic Environmental Impact Statement: Solid Waste Management Plan for Dutchess County*. Prepared for the Dutchess County Resource Recovery Agency. Poughkeepsie, New York: March, 1991.

San Francisco, Office of the Chief Administrative Officer (Amy Perlmutter and Maria Hon). "San Francisco's Comprehensive Reycling Program." *Public Works Magazine*. July 1991.

San Francisco Recycling Program. *Environmental Shopping Guide*. 1990.

Seattle Solid Waste Utility. *Use It Again, Seattle!* 1992.

Seattle Solid Waste Utility (Lisa A. Skumatz and Cabell Breckinridge). *Variable Rates in Solid Waste: Handbook for Solid Waste Officials: Volume I - Executive Summary*. June 1990.

Seattle Solid Waste Utility. *On The Road to Recovery: Seattle's Integrated Solid Waste Management Plan*. August 1989.

Seattle Solid Waste Utility (Lisa A. Skumatz, Ph.D.). *Volume-Based Rates in Solid Waste: Seattle's Experience*. 1989.

SJ Berwin & Co. (Michael Rose). "The German Waste Packaging Ordinance: Recycling Uber Alles?" Kent, England: *Warmer Bulletin*. November, 1991.

Smolowe, Jill. "Read This!!!" *Time*. November 26, 1990.

Specter, Michael. "Dinkins' Role in Sanitation Is Faulted." *The New York Times*. January 18, 1992.

Tellus Institute. *Existing and Future Solid Waste Management Systems in the RPA Region*. Prepared for Regional Plan Association. Boston: March 1992.

Tellus Institute (John Stutz and Gary Prince). *A Statistical Profile of New York City for Solid Waste Management Planning*. Prepared for The New York City Department of Sanitation. Boston: May 17, 1991.

Tellus Institute. *Disposal Cost Fee Study - Final Report*. Prepared for the California Integrated Waste Management Board. Boston: February 15, 1991.

Tellus Institute, Recourse Systems, and James C. Anderson Associates. *Solid Waste Management Issues and Options for New York City: A Report to the Manhattan Citizen's Advisory Committee on Solid Waste*. March 1990.

Trombly, Jeanne. "Airline Recycling Takes Off," *Resource Recycling*. February 1991.

US Environmental Protection Agency (J. Winston Porter). *The Solid Waste Dilemma: An Agenda for Action*. Washington, DC: February 1989.

US Environmental Protection Agency. *Bibliography of Municipal Solid Waste Management Alternatives*. Washington, DC: August 1989.

US Environmental Protection Agency. *Characterization of Municipal Solid Waste in the United States: 1990 Update*. Washington, DC: June 1990.

US Environmental Protection Agency. *The Environmental Consumer's Handbook*. Washington, DC: October 1990.

US Environmental Protection Agency. *EPA's Existing Chemicals Program: An Overview*. Washington, DC: 1991.

US Environmental Protection Agency. *Promoting Source Reduction and Recyclability in the Marketplace*. Washington, DC: September 1989.

US Environmental Protection Agency. *Report to Congress: Solid Waste Disposal in the United States, Volume II*. Washington, DC: 1988.

US Environmental Protection Agency. *Summary of Source Reduction Activities in 25 Selected States*. Washington, DC: May 1991.

US Environmental Protection Agency. *A Survey of Household Hazardous Wastes and Related Collection Programs*. Washington, DC: 1986.

US Environmental Protection Agency. *Unit Pricing: Providing an Incentive to Reduce Municipal Solid Waste*. Washington, DC: February 1991.

US General Accounting Office. *Solid Waste: Trade-offs Involved in Beverage Container Deposit Legislation*. Washington, DC: November 1990.

US International Trade Commission. *Synthetic Organic Chemicals: US Production and Sales*. Washington, DC: 1989.

University of Arizona, Environmental Systems Monitoring Laboratory, *Characterization of Household Hazardous Waste from Marin County, California, and New Orleans Louisiana*. 1987.

Warmer Bulletin. November 1991.

Waste Age. November 1990.

Waste Management, Inc. and Piper & Marbury, Inc. *Waste Reduction: Policy & Practice*. New York: 1990.

Waste-Tech. *The New York City Medical Waste Management Study, Task 4 Final Report: The New York City Medical Waste Management Plan*. June 24, 1991.

Westchester County and Southern Westchester Board of Cooperative Educational Services. *Recycling - How it Fits in Your Curriculum*. New York: 1991.

Wisconsin Department of Natural Resources. *Recycling Study Guides*. 1990.

World Wildlife Fund & The Conservation Foundation. *Getting at the Source: Strategies for Reducing Municipal Solid Waste*. Washington, DC: 1991.

Worldwatch Institute (Sandra Postel). *Worldwatch Paper #79: Defusing the Toxics Threat: Controlling Pesticides and Industrial Waste*. Washington, DC: 1987.

Worldwatch Institute (John E. Young). *Worldwatch Paper 101: Discarding the Throwaway Society*. Washington, DC: January 1991.

Zweizig, Miriam, *et al.* "Individual Source Reduction Behavior: A study of the Effect of Environmental and Economic Motivational Information." University of Michigan School of Natural Resources Master's Project. Ann Arbor: June 21, 1991.

Index

About the Authors

Bette K. Fishbein

Bette Fishbein is Director of INFORM's Municipal Solid Waste Program, where her efforts are currently focused on how the government, communities, and businesses can effectively reduce the amount of garbage they generate.

Prior to joining INFORM, she served as Issues Director for the Ravitch mayoral campaign in New York City, where she developed the candidate's policy positions on solid waste and other issues. From 1974 to 1987, Ms. Fishbein was a staff economist at the Institute for Socioeconomic Studies in White Plains, New York. She also worked as a research analyst at the National Bureau of Economics in New York and at Resources for the Future in Washington, DC.

Ms. Fishbein holds a B.A. from Wellesley College where she earned honors in economics and was Phi Beta Kappa.

Caroline Gelb

Caroline Gelb joined INFORM's Municipal Solid Waste Program as a researcher in January 1991. Primarily, she analyzes source reduction initiatives that can be adopted by municipalities and businesses. Ms. Gelb is currently studying corporate initiatives that are being taken to reduce the amount of waste generated by their products and packaging. She is also working on the design of a major source reduction educational campaign for New York City.

Previously, Ms. Gelb served as a policy analyst with the Office of Air Quality at the New York City Department of Environmental Protection.

Ms. Gelb earned her B.A. in international relations and natural resources from the University of Michigan.

INFORM Publications and Membership

Municipal Solid Waste Selected Publications

Case Reopened: Reassessing Refillable Bottles (David Saphire), Spring 1994, ca. 150 pp., $25.

Germany, Garbage, and the Green Dot (Bette K. Fishbein), 1994, 270 pp., $25.

Business Recycling Manual (copublished with Recourse Systems, Inc.), 1991, 202 pp., $85.

Burning Garbage in the US: Practice vs. State of the Art (Marjorie J. Clarke, Maarten de Kadt, Ph.D., and David Saphire), 1991, 288 pp., $47.

Reducing Office Paper Waste (Robert Graff and Bette Fishbein), 1991, 28 pp., $15.

Garbage Management in Japan: Leading the Way (Allen Hershkowitz, Ph.D., and Eugene Salerni, Ph.D.), 1987, 152 pp., $15.

Garbage Burning: Lessons from Europe: Consensus and Controversy in Four European States (Allen Hershkowitz, Ph.D.), 1986, 64 pp., $9.95.

Forthcoming Publications on Municipal Solid Waste
Clean Products Study 2: Reusable Shipping Containers (working title)

Chemical Hazards Prevention Selected Publications

Preventing Industrial Toxic Hazards: A Guide for Communities (Marian Wise and Lauren Kenworthy), 1993, 208 pp., $25.

Environmental Dividends: Cutting More Chemical Wastes (Mark H. Dorfman, Warren R. Muir, Ph.D., and Catherine G. Miller, Ph.D.), 1992, 288 pp., $75.

Tackling Toxics in Everyday Products: A Directory of Organizations (Nancy Lilienthal, Michèle Ascione, and Adam Flint), 1992, 192 pp., $19.95.

Toward a More Informed Public: Recommendations for Improving the Toxics Release Inventory (Jacqueline B. Courteau and Nancy Lilienthal), 1991, 26 pp., $10.

Preventing Pollution Through Technical Assistance: One State's Experience (Mark H. Dorfman and John Riggio), 1990, 72 pp., $15.

Cutting Chemical Wastes: What 29 Organic Chemical Plants Are Doing to Reduce Hazardous Wastes (David J. Sarokin, Warren R. Muir, Ph.D., Catherine G. Miller, Ph.D., and Sebastian R. Sperber), 1986, 548 pp., $47.50.

Energy and Air Quality

Paving the Way to Natural Gas Vehicles (James S. Cannon), 1993, 192 pp., $25.

Other INFORM Publications

For a complete publications list, including materials on land and water conservation and a quarterly newsletter, or for more information, call or write to INFORM.

Sales Information

Payment

Payment, including shipping and handling charges, must be in US funds drawn on a US bank and must accompany all orders. Please make checks payable to INFORM and mail to:

> INFORM, Inc.
> 381 Park Avenue South
> New York, NY 10016-8806

Please include a street address; UPS cannot deliver to a box number.

Shipping Fees

United States:	add $3 for first book + $1 for each additional book (4th class delivery; allow 4-6 weeks)
Canada:	add $5 for first book + $3 each additional book
Foreign/surface:	add $8 for first book + $4 each additional book
Foreign/airmail:	add $20 for first book + $10 each additional book
Outside the US:	allow additional shipping time

Priority shipping is higher; please call for charges.

Discount Policy

Booksellers:	20% on 1-4 copies of same title
	30% on 5 or more copies of same title
General bulk:	20% on 5 or more copies of same title

Public interest and
community groups:

	Price:
Books under $10:	no discount
Books $10-$25:	$10
Books $26 and up:	$15

Government, upon request:

Books $45 and under	no discount
Books over $45	$45

Returns

Booksellers may return books, if in saleable condition, for full credit or cash refund up to 6 months from date of invoice. Books must be returned prepaid and include a copy of the invoice or packing list showing invoice number, date, list price, and original discount.

Membership

Individuals provide an important source of support to INFORM and receive the following benefits:

Member ($25):	A one-year subscription to *INFORM Reports*, INFORM's quarterly newsletter.
Friend ($50):	Member's benefits and advance notice of new publications.
Contributor ($100):	Friend's benefits, plus a 10% discount on new INFORM studies.
Supporter ($250):	Friend's benefits, plus a 20% discount on new INFORM studies.
Donor ($500):	Friend's benefits, plus a 30% discount on new INFORM studies.
Associate ($1000):	Friend's benefits, plus a complimentary copy of new INFORM studies.
Benefactor ($5000):	Friend's benefits, plus a complimentary copy of new INFORM studies.

All contributions are tax-deductible.

BOARD OF DIRECTORS